LabTutor

A Friendly Guide to Computer Interfacing and LabView Programming

John K. Eaton and Laura Eaton

New York   Oxford
OXFORD UNIVERSITY PRESS
1995

Oxford University Press

Oxford New York
Athens Auckland Bangkok Bombay
Calcutta Cape Town Dar es Salaam Delhi
Florence Hong Kong Istanbul Karachi
Kuala Lumpur Madras Madrid Melbourne
Mexico City Nairobi Paris Singapore
Taipei Tokyo Toronto

and associated companies in
Berlin Ibadan

Copyright © 1995 by Oxford University Press, Inc.

Published by Oxford University Press, Inc.,
200 Madison Avenue, New York, New York 10016

Oxford is a registered trademark of Oxford University Press

All rights reserved. No part of this publication may be reproduced,
stored in a retrieval system, or transmitted, in any form or by any means,
electronic, mechanical, photocopying, recording, or otherwise,
without the prior permission of Oxford University Press.

Library of Congress Cataloging-in-Publication Data
Eaton, John K.
LabTutor: a friendly guide to computer
interfacing and LabView programming/
John K. Eaton and Laura Eaton.
p. cm. Includes index.
ISBN 0-19-509162-0
1. LabView.
2. Scientific apparatus and instruments—
Computer simulation.
3. Computer graphics.
I. Eaton, Laura. II. Title.
Q185.E28 1995 005.369—dc20 94-23129

9 8 7 6 5 4 3 2 1
Printed in the United States of America
on acid-free paper

Preface

LabTutor was developed to fill the need for a general textbook in computer interfacing while simultaneously providing the option of self-guided learning of the subject. We chose a hypertext presentation of the material as much out of curiosity as necessity. We now believe that the addition of sound and animation to the ordinary text and graphics significantly enhances the understanding of the subject matter. We believe strongly that computer interfacing can only be learned by doing. Therefore, we have designated a specific programming language, LabView, from National Instruments, as the basis for programming examples. We selected LabView because it is relatively easy for students to learn, but also has many advanced features needed by professionals.

We encourage the reader to use the Hypercard version rather than a printed version, and to work from front to back on the first pass through the tutorial. In each section you will learn the basics behind a particular type of interfacing then proceed to exercises that will guide you through using the interface with LabView. It is very important that you work through all the exercises. You will not need the LabView manuals other than "Getting Started" unless you do very advanced applications.

We have received much help from present and former students at Stanford, most notably, Scott Abrahamson, Debora Compton, Jonathan Kulick, Ellen Longmire, Curt Nelson, Chris Rogers, and the hundreds of students who have been guinea pigs in courses using *LabTutor*. The initial development was done under a grant of release time from the Stanford University Fellow's program. Extensive additions were financed by the National Science Foundation under Grant USE-9053617 as part of the Undergraduate Curriculum and Course Development Program. We received extensive support including both equipment and programming assistance from Apple Computer. National Instruments personnel have offered both encouragement and software support throughout the development, and we received generous gifts of Tektronix and Fluke instrumentation.

This book is dedicated to our children Elaine and Diana, our continuing source of inspiration.

Stanford, Calif.   J. K. E., L. I. J. E.
Spring 1994

Contents

Chapter 1 Introduction, 3

Chapter 2 Getting Started with LabTutor, 5

Chapter 3 Introduction to Computer Interfacing, 7

Chapter 4 Interface Boards, 16

    4.1 National Instruments NB-MIO-16, 16
    4.2 National Instruments NB-GPIB, 18

Chapter 5 Learning LabView Basics, 19

Chapter 6 Laboratory Instruments, 21

    6.1 Introduction, 21
    6.2 Cabling Systems, 21
    6.3 Function Generator, 22
    6.4 Oscilloscope, 23
    6.5 Digital Multimeter, 27
    6.6 Interface Tester, 32

Chapter 7 Digital-to-Analog Conversion, 34

    7.1 Introduction to D-to-A Conversion, 34
    7.2 DAC Hardware Overview, 36
    7.3 Selection Criteria, 39
    7.4 DAC Software Overview, 41
    7.5 LabView Control of the DAC, 43
    7.6 Hardware Exercises, 48

Chapter 8 Analog-to-Digital Conversion, 50

    8.1 Introduction to A-to-D Conversion, 50
    8.2 ADC Hardware Overview, 51
    8.3 ADC Auxiliary Hardware, 54
    8.4 Selection Criteria, 56
    8.5 ADC Software Overview, 58
    8.6 LabView Control of the ADC, 58
    8.7 Hardware Exercises, 77

Chapter 9 Parallel Digital Interfacing, 82

- *9.1* Introduction to Digital I/O, 82
- *9.2* Digital Interface Hardware, 83
- *9.3* Parallel Interfacing Protocols, 86
- *9.4* Parallel Interfacing with LabView, 89

Chapter 10 The IEEE-488 General Purpose Interface Bus, 94

- *10.1* Introduction to the GPIB, 94
- *10.2* GPIB Hardware, 95
- *10.3* Transmitting Information on the GPIB, 97
- *10.4* Operation the Bus from LabView, 98
- *10.5* GPIB Programming Examples, 100

Chapter 11 Analysis of Sampled Data, 106

- *11.1* Introduction, 106
- *11.2* Random Variables, 108
- *11.3* The Probability Density Function, 111
- *11.4* Mean and Standard Deviation, 117
- *11.5* Higher Moments of the PDF, 128
- *11.6* Correlations, 130
- *11.7* Power Spectrum, 148

Index to the LabTutor VIs, 169

LabTutor

Chapter 1

Introduction

Modern microprocessor technology and powerful personal computers have had an enormous impact on the way laboratory experiments are conducted. Instrumentation systems are now controlled entirely by computer in modern laboratories, and data are acquired directly through plug-in interface boards. Most data processing, display, and storage can be done on the personal computer too. A full laboratory data acquisition and control system, including software, can now be purchased for under $5,000 meaning that the cost of the computer is generally considerably less than that of the sophisticated instruments used in many laboratories.

The use of a laboratory computer does more than make the work more convenient for the experimenter; it usually also improves the quality of the resultant data. Human errors in the reading of instruments and the transcription of data are eliminated. Also, more data samples are usually acquired when a laboratory computer is used allowing a reduction in statistical errors. The nearly immediate processing of the data allows the experimenter to check the results and to catch obvious problems before shutting down the experiment.

More important, the use of a laboratory computer opens up entirely new possibilities. Experiments are now routinely performed that could not be conceived before the advent of the laboratory computer. When the experimental hardware is controlled by the computer, output variables may be sensed and appropriate control signals sent to allow direct interactive control of the experiment.

Experimenters with special requirements and skills have been making use of laboratory computers for many years. However, the computers were expensive, and their effective use often required advanced skills in electrical engineering, computer hardware, and software development. Now, the computer systems are relatively inexpensive, and a scientist or engineer with general laboratory skills can use them effectively. We have reached the point where virtually every worthwhile experiment should be interfaced to a computer. Engineers and scientists should at least understand the capabilities of a modern laboratory computer so that they can specify appropriate experiments. Those of us who work even occasionally in laboratories should have the capability to write our own interface programs and to connect the computer to the experiment.

Two problems that have delayed the more widespread use of PC-based laboratory computer systems are potential users intimidation by electronic devices and the difficulty of learning interfacing software. Recently, a number of software products have been introduced that make programming of external interfaces much easier. Most notable is the LabView program developed by National Instruments and referred to in this tutorial. This graphical programming "language" is much easier to learn and remember than conventional languages and helps the user to identify problems with the interface program.

Even with the modern software, learning computer interfacing can be quite difficult for those with little familiarity with the basic concepts. This is where *LabTutor* comes in. It is designed to help you learn how computer interfaces work, what is possible to do, how to select new equipment, and how to use LabView. Each chapter starts by covering the basics of a different type of interfacing, then covers how to use LabView for relatively simple tasks. Exercises and examples help you learn some of the more difficult LabView techniques. We strongly encourage you to take a hands-on approach to learning about computer interfacing. You can learn to use a computer only if you actually use it. *LabTutor* should make it relatively easy to learn enough computer interfacing to set up your own computerized laboratory. Once you have mastered the skills taught here, you will be ready to delve into more sophisticated interfacing for special applications in your own field.

LabTutor has been designed and extensively tested for use as a self-guided tutorial and as a classroom textbook. If you are using *LabTutor* to learn computer interfacing on your own, we suggest that you set up an external connection box similar to the one described in Chapter 4 and an interface tester as described in Chapter 6. An oscilloscope, a digital multimeter, and a function generator will also be very useful. If you are learning interfacing in an established laboratory, some of the equipment will not be precisely as described here. However, you will find that most of what is written here generalizes quite well to other hardware.

LabTutor can be used as a textbook in either its hard copy or its HyperCard stack form. The computer version is somewhat more complete because it includes animation and sound, which are not available in hard copy. *LabTutor* can also be used as a reference work because it is based on HyperCard. The various levels of menus available allow you to move quickly through the stack to find the information you need. Also, you can use HyperCard's ability to search rapidly through a stack in order to find any available information. Thus, *LabTutor* can be treated as a reference book with a very detailed index.

If you are completely unfamiliar with computerized data acquisition and control, we suggest that you read through the first several sections running the animations until you get to the exercises in the Digital-to-Analog Conversion section of the stack. At this point you will be ready to send out voltages and will begin to feel in control of the computer. You may then proceed through the later sections working the examples as you go.

Alternatively, you might treat *LabTutor* as you would the manual for software you have just purchased. You can flip through the cards looking for something you want to try, then use the HyperCard search capabilities to find the information you need to complete the exercises. You will probably find that you need to work through the introductory sections of the LabView manual before you can make much progress with the exercises.

Finally, if you are already familiar with computerized data acquisition and control you should jump directly to the Introduction to LabView section, then proceed to the subsections LabView Control of the DAC, LabView Control of the ADC, etc. You can then use *LabTutor* as a reference if you have problems when you are developing your own software.

Chapter 2

Getting Started with LabTutor

To use all of the features of *LabTutor,* you need an Apple Macintosh personal computer with the HyperCard software installed. You also need two interface boards from National Instruments, an NB-MIO-16 Multifunction I/O Board and an NB-GPIB interface board. These boards are easy to install following the instructions shipped with the boards.

All of the exercises require that you have the LabView software. This software is not part of *LabTutor* and must be purchased separately from National Instruments. If you are using *LabTutor* as part of a course, your instructor will have installed the interface boards and the LabView software.

You will need a set of laboratory instruments, including an oscilloscope, a function generator, a power supply, and at least one instrument controllable by the IEEE-488 Bus. Specific instruments are listed in the Instrumentation section, but you can use any instruments you have available.

HyperCard is a flexible software program that allows you to search quickly through a variety of data that may be text, illustrations, digitized photographs, video sequences, or sound. The data are stored on cards, which are analogous to file cards. You use HyperCard to move through the stack of cards to find the information you are seeking. Run the HyperCard tutorial to learn how to use the software or just follow the instructions below.

To read *LabTutor*, you should first install it on your hard disk, following the instructions shipped with your copy. HyperCard is a little too slow when run from a floppy disk. Double-clicking on the *LabTutor* icon will start HyperCard and load the *LabTutor* stack. You will be looking at the first card of the *LabTutor* stack. Each card of the stack looks similar to this page, with a title at the top and the forward and backward arrows at the bottom center. A menu bar appears at the top of the screen.

There are several ways to move around the *LabTutor* stack. The simplest is to click on either the forward or the backward arrow at the bottom. These buttons cause HyperCard to flip to the next or to the previous card respectively. Try them and see how rapidly you can move from card to card. Most *LabTutor* cards include a button at the bottom of the card to move to the section menu. The section menu includes a button to get back to the main menu. Menus are a set of labeled buttons. Clicking on the appropriate button moves you directly to that section of the stack. Note that buttons momentarily turn black to indicate they have been clicked.

The menu bar at the top of your screen should have three headings: File, Edit, and Go. Point the mouse to any item on the menu bar and hold down the button to see the menu. The Go menu in Figure 2.2 provides useful functions for moving around the stack. You can select any of these functions using the menu or by typing the indicated command character. Typing either ⌘M or ⌘F brings up a message box such as the one shown in Figure 2.2. If you use the ⌘F, the message box will come up with the command find and an empty set of quotation marks. Type the string you want to locate between the quotes. Typing return will move you to the next occurrence of the string. Find may also locate substrings. For example, if you type the string "to start," Find will locate every occurrence of the word "to." To avoid this, type: find whole "to start." Try this and see how it works. To return to this card, use the Go-Recent command from the Go menu. You will have to look fairly carefully, since the display of miniature cards is not necessarily in the order that you looked at them.

Back	⌘~	Returns to the last card displayed.
Home	⌘H	Goes to the Home card
Help	⌘?	Brings up the Help file. Click Exit to get out of Help.
Recent	⌘R	Shows miniature pictures of the last several cards displayed. Click on any one of them to go back to that card. Try it!
First	⌘1	Goes to the first card in the stack.
Prev	⌘2	Goes to the previous card in the stack.
Next	⌘3	Goes to the next card in the stack.
Last	⌘4	Goes to the last card in the stack.
Find	⌘F	Brings up the Find dialog box described on the next card.
Message	⌘M	Brings up the Message box described on the next card.

Fig. 2.1 Hypercard commands

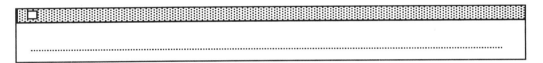

Fig. 2.2 Find window

Chapter 3

Introduction to Computer Interfacing

A laboratory computer must serve a variety of functions, including acquiring data directly from transducers, issuing control signals to various actuators, communicating with smart instruments, processing results, and storing data. The focus of *LabTutor* is on the connection or interfacing of the computer to external devices, such as transducers, motors, light arrays, digital circuitry, and intelligent instruments.

The laboratory computer must be capable of reading in or sending out several different types of signals. The most common signals in laboratory practice are analog voltages. These analog voltages must be converted to digital numbers when stored in the computer. A second type of signals is digital logic signals which are not necessarily directly compatible with the digital signals in the computer. Finally, special purpose digital interfaces are required to connect the computer to a wide range of laboratory instruments that have their own on-board intelligence. The laboratory computer must coordinate the activity of the various instruments in the lab, so it needs a way to send instructions to and to receive data from the intelligent instruments.

Many transducers, including thermocouples, pressure transducers, strain gauges, accelerometers, and flow meters, produce low-voltage analog signals. Other actuators, such as DC motors and variable intensity light sources, require analog control voltages. The computer uses two special devices to send out and read in analog voltages: the digital-to-analog converter and the analog-to-digital converter. These two devices allow the computer to be connected directly to many of the most common transducers used in the laboratory without any additional hardware. An analog voltage signal source may be connected directly to the A-to-D converter as long as the signal level is fairly low, typically less than 10 volts. The D-to-A converter can be connected to control actuators providing the actuator doesn't draw too much current from the D-to-A (typically less than about 20 milliamps). D-to-A and A-to-D Converters are described in detail in Chapters 7 and 8 of *LabTutor* respectively. This may be a good place for the new user to start.

Some sensors and actuators are inherently digital and are therefore most appropriately connected to the computer via a direct digital link. Examples include digital position encoders, frequency counters, stepper motors, relays, solenoids, switches, and lights. A parallel digital interface is used to connect these devices to the computer. The parallel digital interface uses TTL (transistor-transistor logic) signal levels, which are common to many other devices. The interface may be connected directly to any device using TTL signal levels but requires special circuitry if any other type of digital signal sources is to be connected.

Another type of digital interface is the General Purpose Interface Bus (GPIB) used for communication between intelligent instruments and the computer. If you have unusual data acquisition requirements such as very high acquisition rates or very high/low voltages, you can do the actual measurement with an intelligent instrument and then pass the data to the computer over the GPIB. The computer maintains central control over the entire experiment but delegates the actual measurement to a more well-suited instrument.

The next few pages describe the basic structure of a typical computer system. Some knowledge of the internal workings of the computer is helpful in understanding various

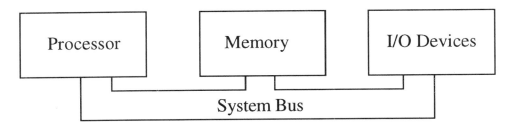

Fig. 3.1 Computer hardware

interfacing issues, but this material is not really critical to successful computer interfacing, so you may skip the rest of this chapter if you are not interested.

A computer may be divided into four logical elements as sketched in Figure 3.1, a processor, a memory, a set of input and output devices, and a system bus connecting all the components. The specific hardware used depends on the type of computer, but every computer has these basic components. A fifth part of the computer that cannot be ignored is the software, or programs of instructions that allow the computer to perform useful tasks.

The processor, or central processing unit, is the central element of the computer. It interprets the sequence of instructions that make up a computer program, performs the actual data manipulations, and sends control signals to the other parts of the computer. Data manipulation may include performing arithmetic and logical operations such as add and compare or storing and retrieving data from memory. The processor must control all of the other functions of the computer. For example, if you wish to send out a voltage signal through the D-to-A converter, the processor must send control signals to set up and initiate the operation.

Modern processors are immensely complicated integrated circuits called microprocessors each fabricated on a single piece of silicon. Most personal computers used in the laboratory are based on one of two families of microprocessors: the Motorola 68000 series and the Intel 80386, 80486, Pentium series. Apple Macintosh computers use the Motorola microprocessors, while IBM-compatible computers use the Intel products. As of 1995, these processors typically could perform around 10-20 million instructions per second.

The memory is the short-term storage area, as distinct from archival storage devices such as magnetic or optical disks. The memory stores the program that is presently active as a sequence of instructions and also stores the data for that program. When the processor executes the program, it gets its instructions sequentially from the memory and moves data in and out of memory as necessary. Logically the memory is configured as a set of storage cells, each having a unique address as illustrated in Figure 3.2. Each cell contains one byte of information; a byte consisting of eight binary digits or bits. The processor may read information from or store information to any one of the cells at any time. Therefore, the memory is called a random access memory, or RAM. Physically, the memory consists of a set of small silicon chips, usually mounted on a small circuit board that is plugged into a socket on the main computer board. RAM memory chips have been the focus of intense international competition, leading to the low memory prices and large memories we enjoy today. A typical personal computer may have several million memory cells or megabytes of storage.

The system bus is the communication pathway shared by all of the computer's components. It is called a bus because it can be used by many different devices to transport data throughout the computer system. It consists of three sub-buses called the address bus, the data bus, and a set of control lines. Physically the bus consists of a large number of parallel conductors on a printed circuit board. Each can carry one bit of information. That

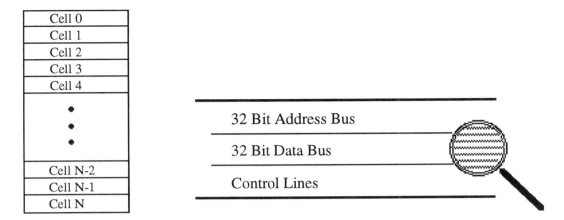

Fig. 3.2 Storage cells   *Fig. 3.3 Basic system bus*

is, each conductor may carry either a high or low voltage signifying either a logical 1 or a logical 0. External devices like a multifunction card used for external interfacing are connected to all of the bus lines by plugging the card into the connector in the computer.

The bus illustrated in Figure 3.3, containing 32 address lines and 32 data lines, is typical of modern personal computers like the Macintosh line of computers. The 32-bit address bus shown can be used to identify up to 2^{32} or more than 4 billion memory cells. Some of the available addresses are used for other devices as discussed below, but this is still a very large number. Present generation personal computers usually have no more than 64 megabytes of memory, but the system bus designers have planned ahead for future systems that may require much larger memories.

The 32-bit data bus can be used to transmit four bytes of information simultaneously. Some devices have the ability to only communicate one or two bytes of data at a time. Such devices would be connected to only 8 or 16 of the 32 data lines.

The control lines serve a number of different functions. Some are used by the processor to control the activity of the memory and the I/O devices. Others are used by the I/O devices to signal the processor when they need attention. The uses of the control lines to control memory reads and writes and I/O device control will be discussed below.

The processor must frequently read instructions and data from the memory and write data back into the memory. This is done using the address and data buses and two control lines as shown in Figure 3.4. To read the contents of a memory cell, the processor must set the address lines to the address of the appropriate cell and then raise the Memory Read line momentarily. The memory sees this signal and after a short delay (called the memory latency) places the contents of the memory cell onto the data bus. The processor may then copy the contents of the data bus into one of its internal registers.

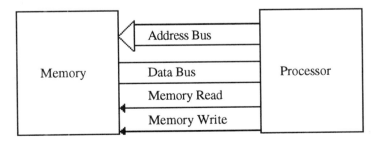

Fig. 3.4 Basic memory read/write

To store data in a memory cell, the processor must place the address of the cell on the address bus and the data to be stored on the data bus. The processor then pulses the Memory Write line and the data are stored by the memory.

The actual interaction between the memory and the processor can be somewhat more complicated than indicated here. This is because of special memory organizations that have been developed to speed up computer systems. The main memory typically has a latency time of around 50 nanoseconds. While this seems fast in terms of everyday experience, it actually is slow relative to the fastest microprocessors now available. If the processor had to wait the full latency each time it attempts to fetch an instruction from memory it would be slowed considerably. Cache memories were developed to alleviate this problem. The cache memory is a small random access memory that is faster than the main memory. Some of the program instructions are stored in the cache. If the next instruction needed is in the cache then the processor can access it very quickly. New parts of the program are brought into the cache as needed, so in most instances the next instruction is available in the cache. The algorithms used to control the cache allocation are difficult to understand. Fortunately, the operation of the cache is completely transparent to even advanced users of laboratory computers.

We now consider a simple computer consisting of only a processor and a memory connected via the system bus. We assume that a program consisting of a sequence of instructions has been stored in the memory. To execute the program, the processor must fetch the first instruction from the memory using the address and data buses. The processor then interprets the instruction and performs the desired operation. In many cases the operation requires that the processor access the memory to store or retrieve data. For example, a possible operation would be to bring in two numbers from memory, add them together, and store the result back in the memory. The instruction must contain enough information for the processor to determine the memory address where the data is stored.

When the processor completes one operation, it brings in the next instruction in the sequence and begins the process anew. We see that the computer code that we write must eventually be decoded into a series of simple instructions that are executed in sequence by the processor.

The simple computer consisting of a processor, a memory, and a system bus can execute programs, but it can't do anything useful because it has no way to communicate with the external world. To do even the basic tasks of loading new programs and interacting with the user the computer must have input/output (I/O) devices. The term *I/O devices* actually encompasses everything you normally associate with a computer other than the processor, memory, and bus. This includes the keyboard, monitor, mouse, disk drives, and other normal accessories, along with the more unusual equipment associated with laboratory computers such as an A-to-D converter or a GPIB interface.

The I/O devices are controlled via the system bus by the processor. Numbers characters and other types of data are passed back and forth across the bus just as they are between the processor and the memory. However, there are a few special techniques and devices that facilitate the interaction between the processor, the memory, and the I/O devices. We will illustrate these techniques by reference to a hypothetical data acquisition card plugged into our computer's bus.

Our simple data acquisition card is illustrated schematically in Figure 3.5. It includes an A-to-D converter, a multiplexer to allow input signals from several different sources, a bus interface unit, and control logic to control the other components of the card. The interface is built on a printed circuit board that plugs into one of the computer's bus connectors, giving it access to all the bus signals and electrical power. The system has several different modes of operation. It can be used to acquire a single voltage sample from one of the input channels and then pass the digital number to the processor. Alternatively, it may be used to acquire a set of voltage samples at regular time intervals. The control logic contains a timer which is used to set the time interval between samples. Finally, the system may be set up to sample the voltage on each of its four inputs sequentially.

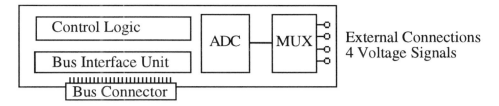

Fig. 3.5 Hypothetical data acquisition card

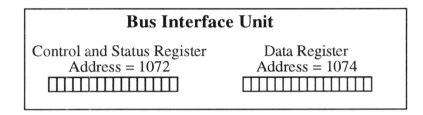

Fig. 3.6 Bus interface unit

Looking into the Bus Interface Unit (Figure 3.6) a little closer, we see that this simple card has two registers, the Control and Status Register and the Data Register. Each of these registers is a 16 bit storage cell that can be accessed using the bus. Each of the registers also has a unique address, which is set when you install the card in the computer. The addresses shown here are imaginary and don't relate to any real system. The addresses must be different from the address of any memory cell or the registers in any other I/O device. The processor may read either register by placing its address on the address bus and then using a control line called I/O Read, which is analogous to the Memory Read Control line. When the Bus Interface Unit sees the address of one of its registers and the I/O Read line is strobed, it places the contents of the register onto the data bus. Similarly, the processor may write into either register.

We are now ready to describe the sequence of operations that occur when you program your computer to acquire a single voltage sample. The processor must first set the card up in the appropriate mode. It does this by writing a control word into the Control and Status Register. This control word is a code that indicates the mode of operation (single sample) and the channel that is to be sampled. Once the control word is received, the multiplexer is switched to the appropriate channel and the A-to-D converter performs the conversion. The resulting digital number is loaded into the Data Register, and one bit of the Control and Status Register is changed to indicate that the conversion is complete. For a typical A-to-D converter, this operation takes about ten microseconds. Once the processor knows that the conversion is complete, it sets the address lines to address 1074 and strobes I/O Read to bring the data sample in for analysis or storage.

One question that should occur to you is: How does the processor know when the conversion is complete? Note that the processor could execute quite a few instructions in the time it takes the interface card to acquire the voltage sample. What does the processor do during this time?

One way for the processor to find out if the analog-to-digital conversion is complete is to read the Control and Status Register which includes a bit to indicate when the conversion is complete. The processor reads the register and performs a logical operation to determine if the completion bit has changed. This sequence of operations is repeated until the data acquisition is complete. The processor then reads the Data Register to bring the voltage sample in for use by the program. Alternatively, the programmer could find out exactly

12 Introduction to Computer Interfacing

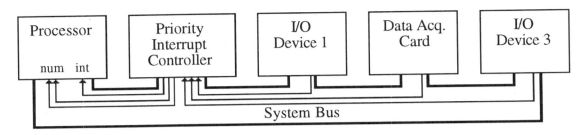

Fig. 3.7 Interrupt system

how long the conversion takes. The program would then command the processor to wait that amount of time before reading the Data Register. This latter type of program is inflexible and unreliable and therefore is not used.

Either solution described above ties up the processor during the entire time the data acquisition is performed. However, the processor could perform many operations in this time if it were freed from monitoring the data acquisition card. For example, the processor could be using this time to update a graphics display, respond to a keystroke, or pass previously acquired data to memory or a disk file. The solution is for the data acquisition card to signal the processor when it has completed using an interrupt system.

The interrupt system allows the data acquisition card to signal the processor directly when it has completed a conversion. The sketch in Figure 3.7 shows that our data acquisition card is just one of the I/O devices connected to the system bus. Each of these I/O devices has a special one-bit output with which it will signal an interrupt. That output is connected via one of the control lines in the bus to a priority interrupt controller. When our data acquisition card completes its conversion and is ready for service, it sends an interrupt signal. The priority interrupt controller sees that interrupt and issues an interrupt signal to the processor through the line labeled "int." Simultaneously, the priority interrupt controller tells the processor which device has interrupted, using the set of lines labeled "num".

When the processor sees the interrupt signal, it completes the instruction it is executing (Inst N in Figure 3.8) and then jumps to an interrupt service subroutine. This subroutine is a special program that is loaded into the computer's memory when you begin to use the data acquisition card. Each I/O device usually has its own interrupt service subroutine. The subroutine first stores the processor's current work and then services the device. In our case, the processor must read the data sample from the card's data register and possibly write out new information to the Control and Status Register. The final instruction of the interrupt service subroutine is a jump back to the main program.

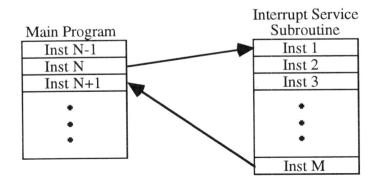

Fig. 3.8 Execution of an interrupt subroutine

The I/O devices are ranked in terms of relative priority for service from the processor. High-priority devices are typically devices with high data transfer rates, such as disk drives or very-high-speed analog-to-digital converters. Low-priority devices are ones that can wait a few milliseconds to be serviced, such as the keyboard. The priority interrupt controller determines if a given device will be able to interrupt an interrupt service routine that is already running. For example, assume that a high-speed data acquisition card has issued an interrupt and the processor is executing its service subroutine. If you press a key, an interrupt will be sent to the priority interrupt controller. However, the priority interrupt controller, seeing that the new interrupt is coming from a low-priority device, will not interrupt the processor until execution of the higher-priority service subroutine is complete.

Fortunately, almost all of this is transparent to the user. You may have to specify an interrupt priority level or an interrupt channel when you install a new interface board in your computer. If so, the instructions that come with the board should explain clearly how to do this. Usually, the board comes with installation software that will take care of this for you.

We next consider a new situation in which we want to acquire a set of 1,000 voltage samples using our data acquisition card rather than a single sample as we've discussed so far. Normally, we would want to acquire these samples at regular time intervals and store them in an array in the computer's memory. There are several different ways in which this can be done, and we need to understand the differences clearly.

Once again, the operation is initiated when the processor writes a code into the Control and Status Register. This code indicates the mode of operation (multiple sample), the input channel to use, the number of samples, and the time interval between samples. On receiving this code, the board control logic initiates the first A-to-D conversion. In a normal interface board, there is very little on-board storage. Therefore, the 1,000 data samples cannot be stored there and must be passed one at a time to the memory. In the simplest mode, the board interrupts the processor each time it has acquired a sample. The processor then reads the data sample over the bus and writes a new value to the Control and Status Register to clear the Data Register for the next sample. Finally, the processor transfers the sample to memory again using the bus.

The data transfer from the data acquisition card to the processor and then to the memory is a two step process, as illustrated in Figure 3.9. The same data sample passes through the bus twice. While this seems wasteful, it is the way that most of the data transfer from data acquisition cards to memory is done. However, there are some cases where this is not possible. If the data acquisition card can acquire data at a very high rate, there may not be enough time for the two-step process to be completed before the next sample must be transferred. Another case is one in which the processor must be performing other operations during the data acquisition. By tying up the bus and the processor, the data transfer hinders any other operations that are running in parallel with the data acquisition program.

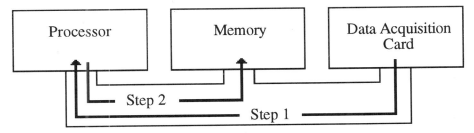

Fig. 3.9 The two steps process required to transfer data from the data acquisition card to memory

14 Introduction to Computer Interfacing

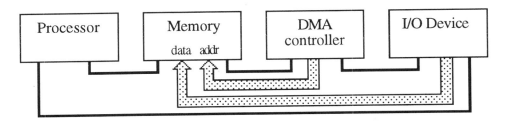

Fig. 3.10 Direct memory access

A different technique called Direct Memory Access or DMA, is available for passing data from an I/O device to the memory. This technique bypasses the step of first passing the data from the I/O device to the processor and instead transfers the data directly from the I/O device to the memory. The organization of a computer system incorporating DMA is sketched in Figure 3.10. A new device, called a DMA controller, is now connected to the bus. During a DMA data transfer from the I/O device to the memory, the datum is placed on the data bus by the I/O device and read by the memory. The memory must know in which memory location to store the datum. The DMA controller supplies the address information, ensuring that all of the data transferred are stored in the appropriate memory locations where the processor can later find them.

Returning now to our example of acquiring a set of 1,000 samples with the A-to-D converter, we will go through the internal steps required. First, the processor must set up the DMA controller by telling it which device will be supplying the data, the number of data samples to be transferred, and the address of the first memory cell to be used. The communication between the processor and the DMA controller is done over the system bus just as if the DMA controller were another I/O device. The processor then sets up the data acquisition card by writing a code to its Control and Status Register, indicating the input channel, the time between samples, the number of samples, and the DMA data transfer mode. The actual voltage sampling then begins. When an A-to-D conversion is complete and the card is ready to transfer the data, it signals the DMA controller using a special control line. The DMA controller waits until a cycle when the bus is not being used, then places the correct address on the address bus and signals the data acquisition card to place the datum on the data bus. The memory then reads the value from the data bus and stores it at the appropriate address. The DMA controller then increments its internal counter so that it is ready to issue the next address.

The process described above is shown in Figure 3.11. We see that, after the initial setup, the entire data acquisition operation leaves the processor free to do other work. For example, the processor may be bringing the data in from the memory and processing it or it may be doing completely unrelated work.

Finally, DMA data transfers may be done from the memory to an I/O device. For example, if a series of voltage levels are to be sent out from a digital-to-analog converter, the array of voltage samples may be stored in memory and passed one at a time to the interface board. Once again, the DMA controller would supply the address information, but in this case the memory would place the data on the data bus and the I/O device would read from it.

Some I/O devices are so fast that even direct memory access is not fast enough to transfer the data across the bus. An example is a flash A-to-D converter, which may acquire data at rates in excess of 100 million samples/sec. In this case, the data acquisition card must have its own dedicated memory (see Figure 3.12) that is fast enough to handle the high transfer rates. During acquisition of a set of voltage samples, the data are stored in the high-speed memory, and there is no interaction with the processor or use of the system bus. After the entire acquisition is complete, an interrupt is sent to the processor to indicate that the conversion is complete. Usually, each cell of the high-speed memory has a unique address and is accessible to the processor. However, in most instances, the data are transferred from the high speed memory to the main memory before processing.

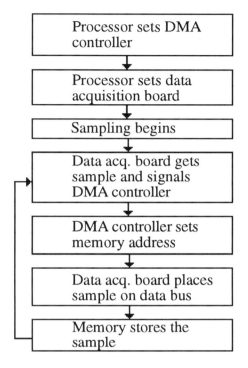

Fig. 3.11 Sequence of operations during a DMA data transfer

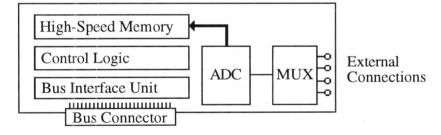

Fig. 3.12 On-board memory

Chapter 4

Interface Boards

In *LabTutor*, we will be using two optional circuit boards to interface the computer to external devices. These boards will allow us to input and output analog voltages, to communicate with digital devices, to count and time events, and to communicate with "smart" instruments.

The boards are:

National Instruments NB-MIO-16   A multifunction interface board including a 12-bit A-to-D converter, two 12-bit D to A converters, an 8-bit digital interface, and three timer/ counters.

National Instruments NB-GPIB   An IEEE-488 controller to communicate with smart instruments on the General Purpose Interface Bus (GPIB).

4.1 National Instruments NB-MIO-16

The NB-MIO-16 is shown in Figure 4.1. It plugs directly into a Macintosh NuBus connector, and its socket for external connections protrudes from the back of the computer. The board can also be connected to the National Instruments Real Time System Integration (RTSI) bus. The board does not have DMA capability by itself but may be connected to a DMA controller via the RTSI bus.

Analog-to-Digital Converter: The board includes a single 12-bit ADC multiplexed to either 16 single-ended inputs or 8 differential inputs. You select which setup you want by moving jumpers. We suggest use of differential inputs for laboratory quality work. The board includes a preamplifier with software selectable gains of either 1, 10, 100, or 500 for an NB-MIO-16L or 1, 2, 4, or 8 for an NB-MIO-16H. The conversion time is nine microseconds, and the maximum sampling rate is 100,000 samples/sec. The board includes a 16-word buffer memory for high-speed conversions. Timing for repetitive sampling is supplied by on-board timers. Alternatively, an external sampling clock may be supplied. The start of sampling may be triggered under software control or by an external trigger signal.

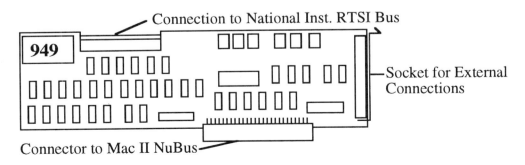

Fig. 4.1 National Instruments NB-MIO-16

Digital-to-Analog Converter: The board includes two 12-bit DACs with a jumper-selectable output range of 0 to 10V or minus 10V to 10V. To duplicate the examples later in the tutorial you should set your board for minus 10V to 10V. The DACs can also be used in the multiplying mode, allowing the user to supply an external reference voltage. The range of the DAC is then equal to the reference voltage. The actual DACs can update their outputs 100,000 times per second, but the NB-MIO-16 board does not include the logic to allow rapid data transfer to the DACs or timed analog outputs. To output a waveform formed by a sequence of voltage samples, the program must transmit each sample individually. This cannot be done very fast, and there is no way to guarantee that the samples will be converted at precise time intervals. The actual conversion rate is dependent on the speed of the computer but it is no more than a few hundred samples/sec.

Digital I/O: The board includes an eight-bit, TTL-compatible, digital I/O port, which can be configured for either digital input or output in four-bit blocks.

Counter/Timers: Three 16-bit counter/timers are available for general purpose use, including event counting, interval timing, and as a frequency source. These counter/timers accept TTL-compatible signals.

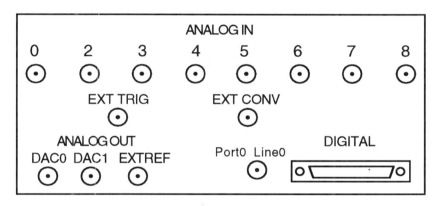

Fig. 4.2 Interface box

Fig. 4.3a Wiring for the analog input section.
The analog input connectors are isolated ground, BNC female connectors. The two connections to each BNC connector are shown in the table. Note that this assumes that you have left the board in the factory setup, that is with differential inputs. The EXT TRIG and EXT CONV connectors are also BNC female connectors.

Input Terminal	NBMIO16 Connector
Ch0+	3
Ch0-	4
Ch1+	5
Ch1+	6
Ch2+	7
Ch2-	8
Ch3+	9
Ch3-	10
Ch4+	11
Ch4-	12
Ch5+	13
Ch5-	14
Ch6+	15
Ch6-	16
Ch7+	17
Ch 7-	18

EXT TRIG + = Pin 38
EXT TRIG - = Pin 33
EXT CONV + = Pin 40
EXT CONV - = Pin 33

18 Interface Boards

Fig. 4.3b   *Wiring for the analog output section.*
There are separate BNC Female connectors for each of the two D-to-A converters. The third connector (EXTREF) allows you to supply an external reference voltage in place of the internal reference source. The ground side for all 3 BNC connectors is connected to Pin 23 of the NB-MIO-16 output connector. The center (positive) pins are connected as shown in the table.

```
     ANALOG OUT
DAC0  DAC1  EXTREF
 ⊙     ⊙     ⊙

DAC0   = Pin 20
DAC1   = Pin 21
EXTREF = Pin 22
```

Fig. 4.3c   *Wiring for the digital section.*
There are 8 digital lines on the NB-MIO-16, which may be configured in groups of 4 as either inputs or outputs. They are connected to 8 pins of the DB25 female connector as shown in the table. Line 0 of Port 0 is also connected to a BNC connector so that the signal may be examined on an oscilloscope.

	NBMIO16 Pin	DB25 Pin
DIO0	25	1
DIO1	27	2
DIO2	29	3
DIO3	31	4
DIO4	26	5
DIO5	28	6
DIO6	30	7
DIO7	32	8
DIG GRND	24	25

The NB-MIO-16 has a rear connector through which all external signals must pass. This connector accepts a standard header-type connector which attaches to a ribbon cable. It is difficult to make signal connections to a ribbon cable so we have connected it to the special interface box shown in Figure 4.2, which allows us to make connections using standard laboratory BNC cables for analog signals and DB25 connections for digital signals. We suggest that you make a similar external connection box before connecting your computer.

4.2 National Instruments NB-GPIB

The NB-GPIB is a low-cost interface for the General Purpose Interface Bus (IEEE-488). It can be used to communicate with up to 14 different instruments through standard GPIB cables. Many instruments from virtually all major instrumentation manufacturers are made with GPIB compatible interfaces.

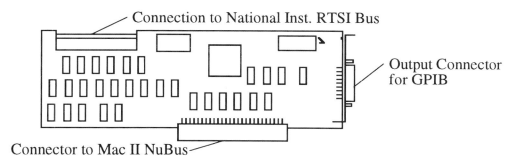

Fig. 4.4 National Instruments NB-GPIB

Chapter 5

Learning LabView Basics

LabView is a programming language and set of subroutines developed by National Instruments for scientific programming and laboratory data acquisition. LabView is different from other programming languages you may have used because you don't type a sequence of steps that are then executed consecutively. Instead, each subprogram and program structure is represented by an icon with various connection points.

All programming in LabView is done graphically. You select the various functions and structures you will need, and they appear as graphical objects on the screen. You do your programming by drawing lines between connection points on the various icons. This is called "wiring" in LabView terminology.

Programs that use one or more LabView functions are called virtual instruments or VIs. Two or more VIs may be combined to make a new virtual instrument. You can make an icon for any new VI you develop and use it in any subsequent VIs.

Each virtual instrument has a Front Panel and a Diagram. The front panel is the place for user interaction. It includes boxes for each input variable and is also used for displaying results. The output display may be numeric, strings, or graphical information. The diagram contains the actual program. The diagram may be very simple or may contain many screens of information.

The LabView software is basically a scientific programming language. It includes all of the basic structures and functions as well as a wide range of virtual instruments to do such functions as vector and matrix algebra, signal processing, and statistical processing. It also includes VIs to control the GPIB. You also need a second set of virtual instruments, called the Data Acquisition VIs to control the multifunction boards such as the NB-MIO-16 we will use. You embed the Data Acquisition VIs within a LabView program to make a complete data acquisition and analysis program. Using these VIs all the complexity of interrupts, DMA, and sample timing should be transparent. You will be able to transfer data into a program from the data acquisition card just as easily as if you were transferring the data from a disk file.

Figure 5.1 shows the diagram for a very simple LabView virtual instrument whose function is to multiply two integer numbers. Values for the two variables, Input A and Input B, are typed into the front panel. The program then computes the product and

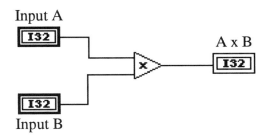

Fig. 5.1 Simple LabView diagram

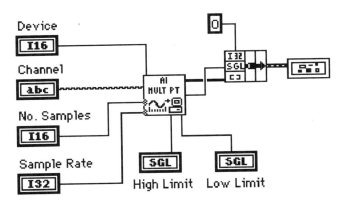

Fig. 5.2 Complicated LabView example

displays it on a front panel indicator labeled A x B. The multiply function appears as a triangle, and the two input variables have been wired to its inputs. The product is wired to the output indicator so that it will be displayed on the front panel.

This example is very simple, and you could probably program it a lot faster in a conventional programming language. However, LabView begins to pay off when the programming gets more complicated. The diagram in Figure 5.2 is from a virtual instrument that acquires a sequence of samples from an A-to-D converter and converts them to voltages. The data are then plotted as input voltage as a function of time. This would be a complicated program in BASIC, FORTRAN, or C. However, we see that the LabView program consists of just a few simple objects connected by wires. One of the best features of LabView is that you can make a new icon for any virtual instrument you develop. That icon can then be used in another, more complicated program.

The manual set for LabView is a little daunting at first glance. Fortunately, you don't have to read through it all to learn to program LabView. You will use most of the manuals only for reference when you develop your own virtual instruments. Much of the information in the manuals is also available via on-line help, so as you gain experience you will find little need for the manuals.

You should start now by sitting at your computer and working through the first six chapters of the LabView Tutorial manual. This will introduce you to the basic structure of LabView and the programming techniques. Some of the examples may seem trivial, but you should work all of them. Many of the programming tricks are introduced in these examples. It will probably take you a few hours to complete, but you then will be a reasonably competent LabView programmer.

Once you've finished those six chapters return to *LabTutor* to learn more about computer interfacing and about programming the Data Acquisition VIs. It will be much easier to learn LabView this way than by trying to just proceed through the manuals.

Chapter 6

Laboratory Instruments

6.1 Introduction

There are a few electronic instruments that are needed in nearly every laboratory. We suggest that at a minimum you have available an oscilloscope, a function generator, and a digital multimeter. These will be needed in order to work all of the exercises in the tutorial, and they are so frequently needed in the laboratory that it is very important for you to become familiar with their use. This chapter will help you learn how to use these instruments, but remember, hands-on practice is very important. The chapter also describes a special interface tester that will help you in testing your LabView programs. Finally, two cabling systems that simplify the instrument connections are described.

6.2 Cabling Systems

Connections between instruments and the computer interface are normally made using BNC cables such as the one illustrated in Figure 6.1. BNC cables are selected because they provide good noise shielding and are easy to connect securely. The signal-return conductor in the coaxial cable is actually a woven flexible wire tube that surrounds the main signal conductor.

We use DB25 cables for digital communications because they are widely available and inexpensive. A DB25 connector is a 25-pin connector with the pins arranged in two rows. We use female connectors on the Interface Tester and on the External Connections Box, so you will need a cable with male connectors on both ends. The connectors are normally joined using 25-conductor ribbon cable. The cable should be wired so that pin 1 of one connector is wired to pin 1 of the other, and so on. If you are not familiar with DB25 connectors, ask for them at your local electronics store.

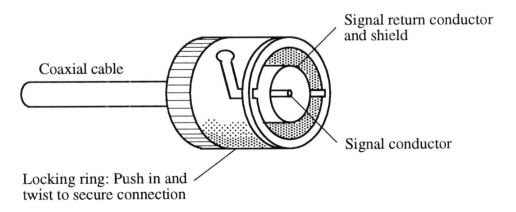

Fig. 6.1 The connector on a BNC cable

22 Laboratory Instruments

6.3 Function Generator

A function generator is used to generate periodic signals. Standard function generators can output a sine wave, a triangle wave, or a square wave at frequencies ranging from a few Hz. up to a few MHz. The function generators usually allow the user to vary the signal amplitude and the mean voltage. Function generators are used to supply test signals, as timing sources, and to power devices requiring AC inputs. This section provides a simplified manual for a commonly used function generator, the Tektronix CFG250. Other function generators are quite similar. You won't be able to test your function generator effectively until you learn to use the oscilloscope.

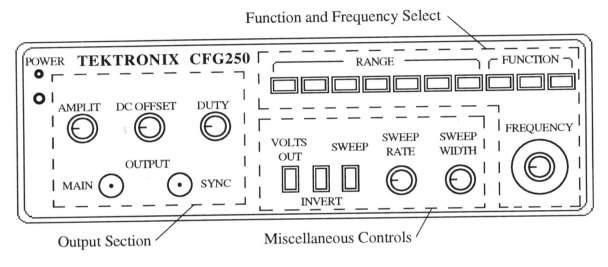

Fig. 6.2 Front panel of the Tektronix CFG250

The front panel of the Tektronix CFG250 is shown in Figure 6.2. The connections and controls are described below for each of the major sections, labeled Output Section, Function and Frequency Select, and Miscellaneous Controls.

Output Section
MAIN OUTPUT: BNC output connector for sine wave, triangle wave, and square wave signals.
SYNC OUTPUT: BNC output connector for TTL signals. The amplitude of the TTL signal is fixed at two volts peak to peak (square wave).
AMPLITUDE: This knob adjusts the output amplitude.
DC OFFSET: This knob changes the DC level and the polarity of the signal at the main output. Pull out the knob to activate. When the knob is pressed in, the signal is centered at zero.
DUTY: This knob changes the time symmetry of the output waveform. When the knob is pushed in, the signal is symmetrical. Pull out the knob to activate.

Function and Frequency Select
FREQUENCY SELECT: The range buttons are used to select the frequency range, and the knob is used to select the exact frequency. To choose a frequency of 16,000 Hertz, push the 10K button and adjust the knob to 1.6.

FUNCTION SELECT: The function select buttons are used to choose the wave form. You can select square wave, triangle wave, or sine wave.

Miscellaneous Controls
VOLTS OUT: Press the button for amplitude range of 0 to 2 Vp-p, open circuit, or 0 to 1 Vp-p into a 50-ohm load. Depress for amplitude range of 0 to 20 Vp-p, in an open circuit, or 0 to 10 Vp-p into a 50 ohm load.
INVERT: Press this button to invert the waveform. This knob will not have any affect unless the DUTY knob has been adjusted. The best way to understand this knob is to connect the function generator to an oscilloscope and try it.
SWEEP: This function generator can be set to sweep through a range of frequencies. Press this button to activate internal sweep and to enable the sweep rate and the sweep width knobs. When the button is set out, the EXTERNAL SWEEP input connector on the rear panel can be used.
SWEEP WIDTH: This knob adjusts the frequency range of the sweep.
SWEEP RATE: This knob adjusts the rate that the function generator sweeps through the frequency range.

Function Generator Exercises
You can try the following exercises after you have learned to use an oscilloscope. First, connect the function generator main output to the oscilloscope input. Set the oscilloscope for a sweep rate of about 1 ms/division. Set the function generator to supply a 1 kHz sine wave. Once you are successfully displaying the signal on your oscilloscope, you should adjust each of the controls in the Output and the Function and Frequency Select sections to understand how they work. The controls in the Miscellaneous Controls section are less important to understand at this point.

The second exercise is to examine how well the function generator can replicate the fast changes required to form a square wave. Set up the function generator to output a 1 kHz square wave varying between -1 and +1 volts. You should see that the signal is not a precise square wave. Increase the frequency to the highest possible for your scope and function generator. Does the waveform change?

6.4 Oscilloscope

An oscilloscope is an instrument that can be used to display a time-varying voltage signal. If the oscilloscope is set up carefully it can be used to measure voltages, but its main function is to provide a qualitative feel for the signal. Although many of the oscilloscope's functions can be mimicked by the computer, generally it is very convenient to have a scope available in the lab to get a preliminary feel for the voltage range and the frequency content of a signal and to examine possible noise sources. For example, high-frequency noise spikes in a signal may contaminate measured statistics but be very difficult to detect by displaying computer-acquired data. It is easy to see the spikes on an oscilloscope trace.

This section provides a brief manual for a very commonly used oscilloscope, the Tektronix 2205. Most oscilloscopes are similar to this, and the wide range of oscilloscopes available from Tektronix all have controls similar to the model 2205. If you have a different oscilloscope available, you should find the manual and explore the scope's use by going to the exercises at the end of this section.

The drawing of the oscilloscope in Figure 6.3 shows the controls divided into sections. The individual controls are labeled in separate smaller figures for each section, and the description is organized by sections.

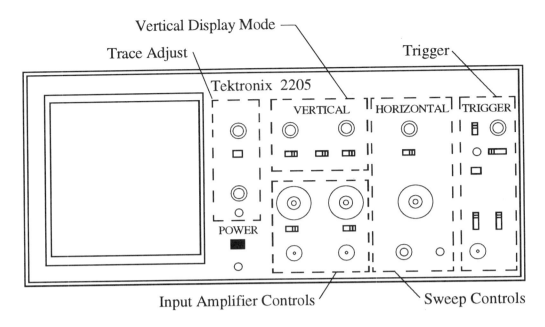

Fig. 6.3 The Tektronix 2205 Oscilloscope

Trace Adjust Controls (Figure 6.4)
BEAM FIND: Use this to find the trace if it is off the screen and you are not sure where it is. Then use position controls to center the trace.
TRACE ROTATION: Turn this knob to make the beam parallel to the center line.
INTENSITY: This knob adjusts the trace brightness.
FOCUS: This knob adjusts the trace thickness.

Input Amplifier Controls (Figure 6.5)
This oscilloscope has two channels, so there are two inputs and an individual set of input amplifier controls for each channel.
INPUT BNC CONNECTOR: There is a standard BNC female input connector for each channel.
INPUT AMPLIFIER GAIN: This knob determines the range of voltages that can be displayed. Adjust it until the scope trace nearly fills the screen. The gain on the two different channels does not have to be the same. The central knob allows the user to adjust the gain between the preset steps.

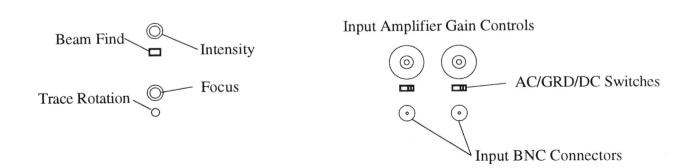

Fig. 6.4 Trace adjust controls   *Fig. 6.5 Input amplifier controls*

AC/GRD/DC: Use AC to see the fluctuating part of the signal. This removes low-frequency oscillations below about 2 Hz from the signal. Use ground to short the input and to adjust the vertical position. Use DC to display the entire signal. You must use this last position to measure voltages with the oscilloscope.

Vertical Display Mode (Figure 6.6)
This set of controls is used to determine which input signal(s) will be displayed and where on the screen they will be displayed. There are two input channels, and you may display either one separately or both together. You can also add or subtract the two signals before displaying the sum as a single trace.
CH 1-BOTH-CH 2: This switch determines which channel will be displayed. Select Ch 1 or Ch 2 to view a channel independently, or select Both to view both of the signals simultaneously.
POSITION: These knobs allow you to adjust the vertical position of each trace. If you need to measure voltages, ground the input, then adjust the Position knob to align the trace with a scale marking.
ADD-ALT-CHOP: The scope actually has only a single beam that is scanned across the screen. In order to display two separate traces it must alternate, drawing first one trace, then the other. The screen phosphor preserves the image for a short time, so you see two continuous traces. ALT-CHOP determines how the scope will operate in dual-trace mode. Use CHOP for low frequency and ALT for higher frequencies. The Add switch position means that the two input signals are added together and displayed as a single trace.
NORM-INVERT: This switch is used to invert the channel 2 signal so that in the Add mode you are displaying Ch1 - Ch2 instead of Ch1 + Ch2.

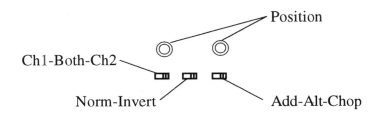

Fig. 6.6 Vertical display mode controls

Sweep Adjust Controls (Figure 6.7)
These controls determine how fast the trace sweeps across the screen and the horizontal position of the start of the trace. You choose the sweep speed depending on the frequency content of your signal.
SEC/DIV: The outer knob selects the horizontal sweep speed. The inner knob is used to select speeds different from the preset values. Normally the inner knob should be turned all the way to the right.
MAG X1 X10: This switch allows you to select the degree of horizontal magnification. Use X10 to increase the sweep speed by a factor of 10.
POSITION: This knob moves the trace horizontally.
This section of our diagram also contains two additional connections which are unrelated to the sweep controls.
PROBE ADJUST: This pin provides a calibration signal, that is a 500 mV peak to peak sine wave at a frequency of approximately 1 kHz.
GROUND: This provides safety ground and direct connection to the ground of a signal source.

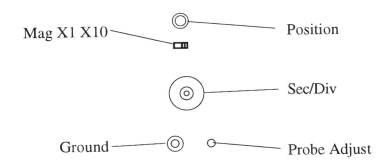

Fig. 6.7 Sweep adjust controls

Triggering Controls (Figure 6.8)
Triggering of the oscilloscope determines when the sweep of the trace begins. If you are observing periodic signals you will want the sweep to begin at the same phase each time so that the waveform will appear to hold still on the screen. Setting up the triggering is the trickiest part of using an oscilloscope. For most normal operations with this scope, setting the Triggering Mode to P-P Auto and the Source to Vertical Mode will work well. The best way to learn the use of all the controls is to experiment. You can't hurt the oscilloscope by doing this, so play as much as you can.

TRIGGERING MODE: Use P-P AUTO for most signals. The sweep will trigger if an appropriate trigger signal is available. Otherwise, the sweep will run continuously. The triggering level is set by the Level control. In NORM mode the scope will trigger only if a trigger signal is received. In SINGLE SWEEP mode the sweep is triggered only once after each reset.
LEVEL: This control determines the threshold voltage at which a sweep is initiated.
SLOPE: This switch determines whether triggering occurs on the rising or falling voltage. For example, if the switch is set to positive and the level set to 3 volts, then the trigger will occur when the input voltage signal rises through a level of 3 volts but will not occur as the voltage passes back down through the same level.
SOURCE: These switches determine what signal will be used for triggering. Normally, one of the input signals CH1 or CH2 is used. The scope can also be set to trigger off the AC supply LINE or from an external synchronizing signal.
RESET BUTTON: This is used to reset the trigger in SINGLE SWEEP mode. The ready light indicates that the scope is awaiting a trigger.
EXT TRIGGER INPUT: This is a BNC female connector used for supplying an external triggering signal.

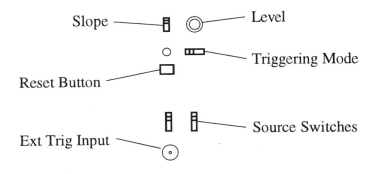

Fig. 6.8 Triggering controls

Oscilloscope Exercise

Connect a BNC cable from the function generator output to the channel 1 input on the scope. Set up the function generator for sine wave output at around 1,000 Hertz. Adjust the scope so that it displays approximately five complete cycles of the sine wave. Use the NORM triggering mode, and observe what happens when you adjust the triggering level and the triggering slope. Sketch the trace in your workbook with the triggering slope positive and negative.

6.5 Digital Multimeter

Digital multimeters are used to measure DC voltage, resistance, and current. Some are also configured to measure AC voltage. In many applications, the multimeter serves the same function as an A-to-D converter. The advantages of using a multimeter are that it usually has a wider range of input voltages, is more accurate, and can easily be set to measure resistance and current. The disadvantage is that the multimeter is considerably slower. It actually uses very high resolution A-to-D converters that operate on a different principle from the successive approximation converters used in computer applications.

Two multimeters, a Tektronix 5120 and a Fluke 8842A, are described here. Both are programmable from either the front panel or remotely over the GPIB. Front-panel operation of the multimeters is described here and computer control via the GPIB is described in Chapter 10.

Tektronix DM5120 Programmable Digital Multimeter

The Tektronix DM5120 shown in Figure 6.9 is an advanced digital multimeter that includes the normal functions of measuring voltage, current, and resistance. It also provides programmable operation for such functions as digital filtering of the voltage samples and automatic subtraction of a null reading. The description of the controls is organized by sections as indicated on the drawing.

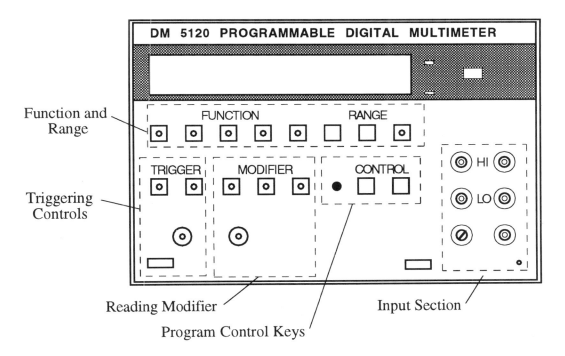

Fig. 6.9 The Tektronix DM5120 Digital Multimeter

28 Laboratory Instruments

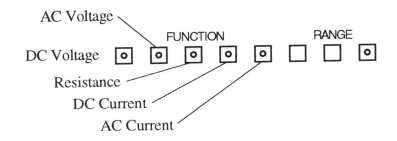

Fig. 6.10 Function and range controls

Function and Range Controls (Figure 6.10)
In normal operating mode these buttons are used to set the function and range of the multimeter. When the multimeter is set in front-panel programming mode, these buttons are part of the keypad used for entering program codes.
DECREMENT: Reduces the range to the next lower (more sensitive) level.
INCREMENT: Increases the range to the next higher (less sensitive) level.
AUTO: Puts the instrument into autorange mode which will automatically select the most sensitive range that does not cause an overrange condition.

Program Control Keys (Figure 6.11)
The multimeter functions can be programmed from the front panel using these control buttons. For example, to record a new Null reading, first press the Program button, then key in the code for null reading (7), and, finally, press enter. The program modes are shown in a small table on the front panel and are described in more detail in the manual.
PROGRAM: Enters the front-panel mode. Using the numeric keys, enter the program number. If you want to terminate an action before it is executed, press PROGRAM.
ENTER: Terminates an entry or confirms that the action performed by the program is the desired action.

Triggering Controls (Figure 6.12)
The triggering controls are used to determine when the multimeter will acquire a measurement.
INT: Selects an internal trigger source, putting the instrument into a continuous triggering mode. This mode operates much like an ordinary multimeter.
EXT/MAN: Selects an external triggering source. When the multimeter is in the external trigger mode, each press of the button will create a trigger and one reading will be taken. Alternatively, the multimeter may be triggered by an external TTL pulse applied at the BNC input.

Fig. 6.11 Program controls

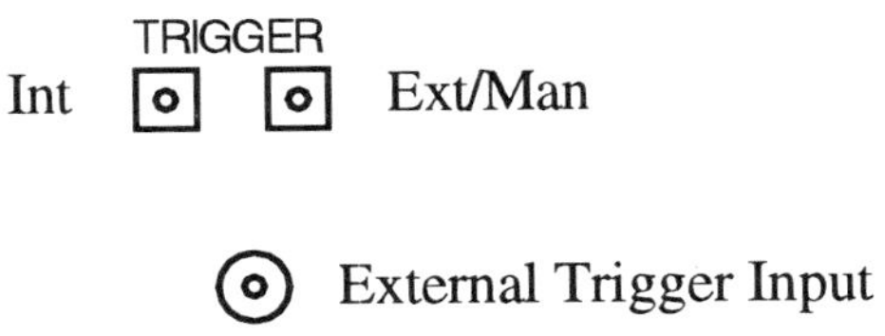

Fig. 6.12 Triggering controls

Reading Modifiers (Figure 6.13)
The multimeter can process the readings in one of several ways before showing them on the display. This processing can be done under front-panel control or can be programmed through the GPIB.
NULL: This mode is used to subtract a number from every reading. For example, it may be used to subtract the zero-pressure voltage output from pressure transducer readings. The null value can be set using program mode 7.
dB: Converts any AC function readings into a decibel (logarithmic) scale. The dB mode allows you to compress a large range of measurements. dB = 20 log (input/reference) where input is the measured AC voltage or current and reference is either 1 V or 1 mA. The reference values can be changed using program mode 9.
FILTER: This button turns on and off the internal weighted average filter. Displayed Value = Last reading + (New reading - Last reading)/Filter value, where the filter value is entered using program mode 8.
READING COMPLETE: Is a TTL-compatible signal indicating completion of a measurement.

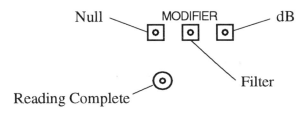

Fig. 6.13 Reading modifier controls

Input Section (Figure 6.14)
HI and LO INPUTS: These input terminals are used for voltage and resistance measurements. You'll need an adapter from double banana lead to BNC to use our normal cabling system.
CURRENT INPUT: Current measurements are made between this input and the center input. Do not attempt to measure currents greater than three amperes.
OHMS SENSE: These inputs are used for four-wire resistance measurements. Four-wire mode is used for very sensitive resistance measurement. See the instrument manual if you need to use this function.
FUSE: Contains the current circuit protection fuse (3 A and 250 V).

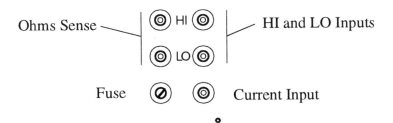

Fig. 6.14 Input section

Fluke 8842A Multimeter

The Fluke 8842 (Figure 6.15) is a very standard multimeter. It has 51/2 digit resolution. It does not do the signal analysis functions of the Tektronix 5120, but it is quite simple to control over the IEEE488 bus.

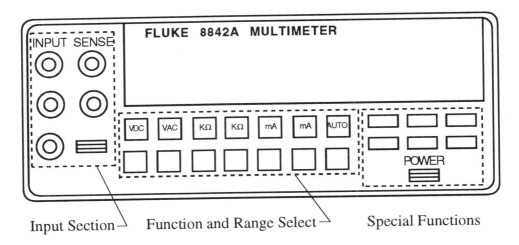

Fig. 6.15 The Fluke 8842A multimeter

Function and Range Controls (Figure 6.16)

The function controls are used to select the type of measurement you would like to make. The use of the buttons is probably obvious. AC measurement capability is an option that is not installed on all 8842As. There are two different buttons for resistance measurement, two-wire and four-wire. Two-wire resistance measurement is the normal measurement of resistance. The value displayed is just the resistance between the two input terminals, so it includes the resistance of the connecting wires. Four-wire resistance measurement is a more accurate technique for measuring the resistance of a component. It requires the use of four wires as illustrated in Figure 6.16. In this case, only the resistance of the component of interest is measured.

RANGE: These buttons are used to select the voltage, current or resistance range of interest.

AUTO: This button enables automatic range selection. If the voltage is fluctuating wildly the range may shift intermittently, making the display difficult to read.

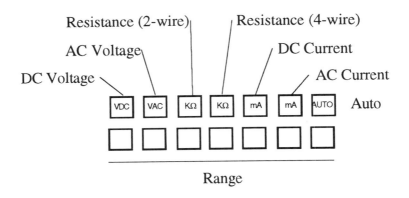

Fig. 6.16 Function and range controls

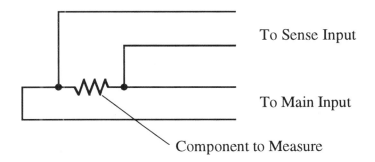

Fig. 6.17 Connection for four-wire resistance measurement

Input Section (Figure 6.18)
HI and LO INPUTS: These are the input terminals used for voltage and resistance measurements. You'll need an adapter from double banana lead to BNC to use our normal cabling system.
2A INPUT: Use this input and the LO input for current measurements.
SENSE: These input terminals are used for four-wire resistance measurements. You need them only for very accurate measurements of resistance.
FRONT/REAR: This switch allows you to select whether you will use the front panel inputs or the separate inputs on the rear panel.

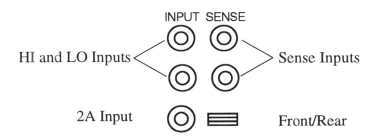

Fig. 6.18 Input section

Special Functions (Figure 6.19)
EX TRIG & TRIG: These buttons allow you to trigger a reading. Use the Ex Trig button to toggle between internal and external triggering. Once in external mode, use Trig to actually trigger the reading.
RATE: This button is used to select the low, medium, or fast reading rate. The display will flash, indicating the reading rate.
OFFSET: This button saves the reading value on the display at the time you push the button. That value is then subtracted from all subsequent readings. You shut this function off by pushing the button again.
SRQ & LOCAL: These two buttons are used in conjunction with the GPIB.

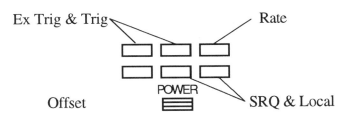

Fig. 6.19 Special functions controls

6.6 Interface Tester

The interface tester is a custom device used for testing the operation of the eight-bit parallel digital interface. It also supplies a DC voltage that can be varied between 0 and 5 volts and is convenient for testing the A-to-D converter and multimeters. A device like this is very useful in a teaching laboratory or if you are starting out on interfacing by yourself. It is constructed from widely available parts, and you need only minimal soldering and electronics assembly skills to build your own. A schematic of the internal electronic components is given in Figure 6.20.

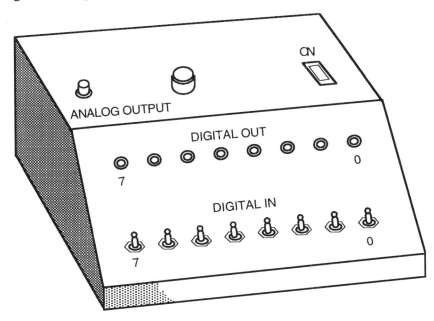

Fig. 6.20 Interface tester

Interface Tester Controls

ANALOG OUTPUT: Provides a convenient analog voltage output to test the A-to-D converter. Vary the voltage between 0 and 5 volts by turning the knob.

DIGITAL OUT: This set of light-emitting diodes (LEDs) is used to test the digital output capabilities of your parallel digital port. Wire the outputs from your port to the Digital Out connector on the back panel. The LED associated with a given bit is on when there is a high state output from the port.

DIGITAL IN: The row of switches is used to send an eight-bit pattern to the digital input port. High corresponds to a 1 and low to a 0. Connect a cable between the Digital In connector on the back panel and the Digital connector on the computer interface box.

The schematics in Figure 6.21 show the internal wiring of the interface tester. You will need a standard 5V power supply to power the various components. The input and output test sections should be connected to separate DB25 connectors on the back panel of the interface tester as shown. If the pin connections shown here are used you will be able to connect a cable directly from the interface tester to the External Connections box described in Chapter 4. The schematic for the Digital In section shows connections to only four of the switches for simplicity. The other four switches should be hooked up in exactly the same way.

Digital Multimeter **33**

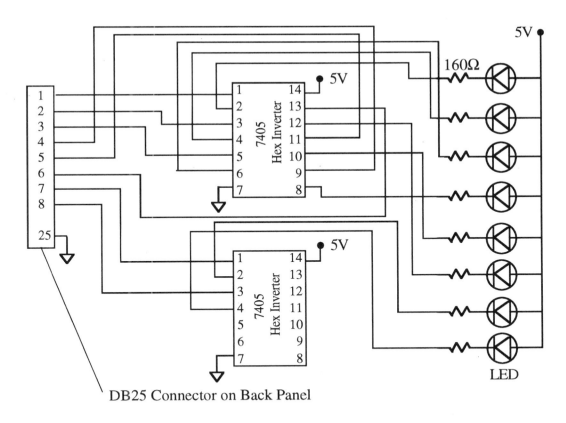

Fig. 6.21a Schematic for the Digital Out section of the interface tester

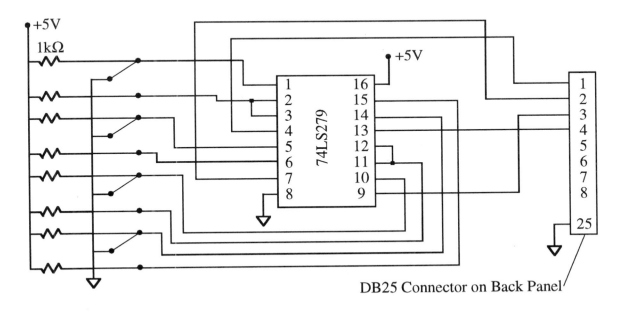

Fig. 6.21b Schematic for the Digital In section of the interface tester

Chapter 7

Digital-to-Analog Conversion

7.1 Introduction to D-to-A Conversion

The function of a digital-to-analog converter (also called a D-to-A converter or a DAC) is to convert a number represented in digital form in the computer to a proportional analog voltage. The schematic symbol in Figure 7.1 shows an n-bit digital input and a single voltage output line. The analog voltage can be connected to an external device that requires a voltage input.

An example of an application of a D-to-A converter is the use of the computer to control the speed of an electric motor. The speed of many motors can be controlled by supplying an input control voltage. In order to command the motor to run at a desired speed, the computer must convert the number representing the desired speed to an actual voltage. The computer can control a wide range of other devices that also accept voltage inputs using the D-to-A converter.

The D-to-A converter is mounted on a printed circuit board plugged into the computer's bus. It receives the digital number from the data bus and converts the number to an analog voltage on command from the board's control circuitry. The voltage is then held until the DAC receives a new digital number and is commanded to convert it. The DAC may be used to produce a stable output voltage or instead may convert a sequence of numbers at regular intervals, creating a time-varying output signal.

A DAC cannot output any arbitrary voltage within its range but instead has a finite number of possible output voltages. For example, a typical DAC may be able to send out voltages between -10 and +10 volts in steps of 5 millivolts. If an output voltage of 1.437 volts is desired it must be rounded to the nearest possible output voltage, or 1.435 volts. The resolution is determined by the number of input bits accepted. Typical D-to-A converters for general purpose use have 8-, 12-, or 16-bit inputs. For example, the National Instruments data acquisition board we are using includes two 12-bit DACs. Converters with lower resolution (fewer input bits) are used in special purpose applications such as digital video.

D-to-A converters may be used to control any device that accepts a low-level voltage signal. Low level generally means less than about 10 volts and 20 milliamps of current. Examples of devices accepting such inputs are motor speed controls, servo control systems, video display units, and oscilloscopes. In many instances, the DAC output can be connected directly to the input of the external device. In other cases, some circuitry is needed between the D-to-A converter output and the device input. For example, DAC output voltages are appropriate to drive small DC motors. However, the DAC cannot

Fig. 7.1 Schematic representation of a digital-to-analog converter

supply sufficient current so a simple amplifier is needed between the DAC output and the motor input.

D-to-A converters are also internal components of many common products. An example is a compact disc player where the data is stored in digital form on the disk. The CD player contains a DAC that converts the digital data to an analog voltage. The resulting voltage is filtered and amplified so that it can produce a sound at the loudspeakers.

The D-to-A converter outputs a voltage that is proportional to the n-bit binary number placed on its input. This transfer function may be represented in equation form as:

$$V_{out} = V_0 + k(b_0 + 2b_1 + 4b_2 + \cdots + 2^{n-1} b_{n-1})$$

V_0 = minimum output voltage

k = slope of the transfer function

b_i = bits of the input number. b_0 is the least significant bit and b_{n-1} is the most significant.

We see that the D-to-A converter has 2^n different possible output voltages and that the step between successive voltage levels is k volts.

We will first examine a typical 12-bit D-to-A converter having an output range of -10 to +10 volts. The 12-bit converter has 4096 possible output states. Then V_o is -10 volts and k is 20 volts/4096 = 0.0049 volts. Therefore, increasing the digital input number by 1 results in an increase of 4.9 millivolts at the output. This smallest possible increment in the output voltage is called 1 least significant bit or 1 LSB.

In the hardware examples at the end of the chapter, we will be examining a hypothetical 4-bit DAC. This DAC has only 16 possible output states. If the converter has the same -10 to +10 volt range, then 1 LSB is 20 volts/16 = 1.25 volts. The plot in Figure 7.2 shows the transfer function for this hypothetical converter. The lowest possible output voltage is -10 volts, which corresponds to an input number of 0. An input number of 1

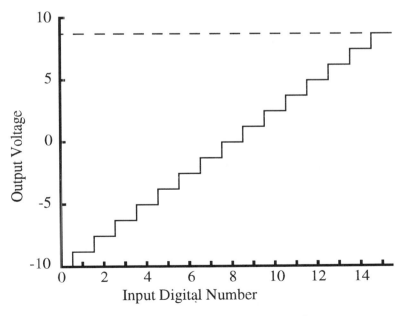

Fig. 7.2 Plot of the transfer function for a four-bit DAC

36 Digital-to-Analog Conversion

corresponds to -10 + 1.25 = -8.75 volts. It is interesting to note that the highest output voltage is -10 + 15(1.25) = 8.75 volts. Therefore, we see that we don't get the full stated range from the DAC. This effect is insignificant with a 12-bit DAC. The actual range for the 12-bit DAC mentioned above is -10 to +9.9951 volts.

7.2 DAC Hardware Overview

There are two types of digital-to-analog converters: current source DACs and voltage source DACs. A current source DAC outputs a current that is proportional to the input digital number, while a voltage source DAC outputs a voltage. The DAC on any general-purpose interface card you purchase will be of the voltage source type, because it is much more common to use a voltage as a control signal than a current. Current source DACs are used in applications requiring very-high-speed conversions or special output conditions. The user of a current source DAC normally adds a high frequency amplifier to convert the current signal to voltage. Fortunately, as a general laboratory user of the computer, you will have no need to do this.

In this section we describe current source DACs first because they are somewhat simpler to understand. A four-bit DAC is used in the illustrative examples. Most DACs have higher resolution; the four-bit DAC was chosen to simplify the diagrams. The extension to a higher resolution should be obvious.

The current source DAC may be thought of as a set of current sources controlled by the bits of the input digital number. Each of the digital inputs (b_0-b_3) controls a semiconductor switch, which opens or closes depending on the value of that bit. The current source corresponding to the least significant bit is the weakest and each successive current source is twice as strong as the previous one. The sum of the currents from all of the controlled sources is the output current that is supplied to the load.

The four-bit current source DAC shown in Figure 7.3 is converting the digital input number 5. The binary representation is 0101, so switches b_0 and b_2 are closed while b_1 and b_3 are open. The total current then flowing out the output is $I_0 + 4I_0 = 5I_0$.

In practice, a current source is just a voltage source and a resistor. A one milliamp current source is shown in Figure 7.4. A 12-bit converter requires current sources with strength from 1 to 2^{12}(= 4096 times I_0). To make an accurate 12-bit DAC then would require us to manufacture a resistor with precisely 4,096 times the resistance of the smallest resistor. This is not impossible but would be very expensive. The DAC could not be manufactured using low-cost integrated circuit technology. The solution is to use a ladder network that uses only two resistance values, as explained below.

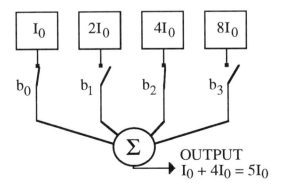

Fig. 7.3 A four-bit current source DAC

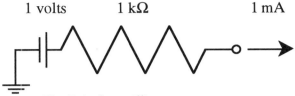

Fig. 7.4 One-milliamp current source

A four-bit current source DAC made using four different resistors is shown in Figure 7.5. The four switches are controlled by the bits b_0 through b_3. If a bit value is zero, the switch is set in the down position; if the value is 1, the switch is set in the up position. Figure out the output current assuming that V_{ref} is 5 volts and R is 1 kΩ. The problem with this DAC is that it requires precision resistors.

The circuit shown in Figure 7.6 is a so-called R-2R network, which is electrically identical to the circuit made using the four different resistors. The key is that it is constructed using only two different resistance values and can be built inexpensively using integrated circuit technology

Most DACs used in computer interfacing applications are voltage source DACs, which work by varying the amplification of a fixed voltage in proportion to the input digital number. The trick is to make an amplifier with a gain that can be precisely controlled over

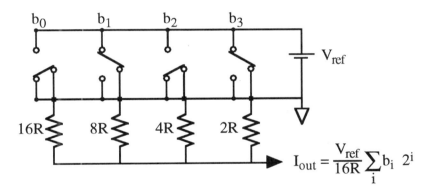

$$I_{out} = \frac{V_{ref}}{16R} \sum_i b_i \, 2^i$$

Fig. 7.5 Four-bit current source DAC made with four different resistors

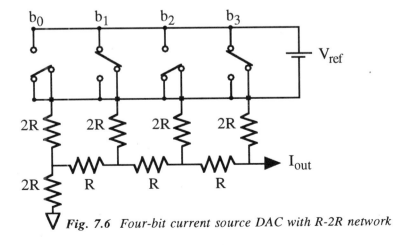

Fig. 7.6 Four-bit current source DAC with R-2R network

38 Digital-to-Analog Conversion

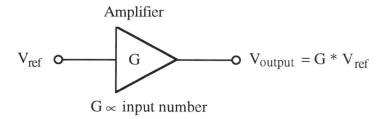

Fig. 7.7 A voltage source DAC constructed from a voltage source and a variable gain amplifier

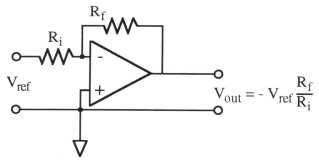

Fig. 7.8 Fixed-gained amplifier constructed using op-amp

a wide range. To do this we use an operational amplifier, which is a simple device used commonly in electronic instrumentation applications. (See Figure 7.7.)

A fixed-gain amplifier built with an op-amp is shown Figure 7.8. The op-amp is the triangle in the middle of the picture. The only other components are two resistors. Note that the gain of this amplifier is just equal to the ratio of the two resistance values. Therefore, a variable gain amplifier may be constructed by having one of the two resistances be variable.

To make a DAC from an op-amp based amplifier we simply make a circuit to vary the input resistance, holding the reference voltage constant. A conceptually simple system is shown in Figure 7.9. The amplifier's input resistor is just the parallel combination of whichever resistors are turned on. For example, if the input number is decimal 10 (1010),

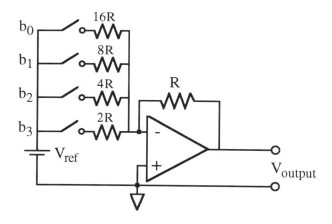

Fig. 7.9 Voltage source DAC with variable resistance

then bits 1 and 3 are turned on. The input resistance is then the parallel combination of 8R and 2R and the output voltage is -0.625 times V_{ref}.

$$R_i = \frac{1}{\frac{1}{8}R + \frac{1}{2}R} = \frac{8}{5}R$$

Output voltage $= -\frac{5}{8} V_{ref}$

Fabricating a DAC with more than a few bits of resolution using the techniques described above would be very difficult. For example, an eight-bit DAC would require manufacturing components with a resistance ratio of precisely 256. This would be virtually impossible in an integrated circuit and extremely expensive in discrete components.

The solution to this problem is again to use an R-2R ladder network which is the exact circuit equivalent of the resistance network shown in Figure 7.9, but is fabricated using resistors of only 2 values, R and 2R. The important point is that R need not be precisely controlled; only the ratio between the two resistance values must be controlled. This can be accomplished readily using integrated circuit technology.

A digital-to-analog conversion subsystem that plugs into a small computer usually includes two or more DACs mounted on an interface card. The actual DACs are relatively inexpensive, but auxiliary hardware is required to make a fully functioning card. Most importantly, the DAC card needs a bus interface unit and control logic to obtain data and control words from the computer's system bus as described in Chapter 3. The bus interface will almost certainly include a system to interrupt the processor and may also include the hardware to allow the DAC to do direct memory access data transfers.

In many cases the DAC will be used to produce a continuously varying voltage signal. In that case, the DAC subsystem will also need a timing source to trigger the DAC at regular intervals when a series of samples are sent out. A high-speed on-board memory may be required if the board is to send out voltage samples at a very high rate. Many cards include a small buffer memory, but for DACs with transfer rates in excess of 1 million samples/sec, a larger on-board memory is usually required.

7.3 Selection Criteria

A wide range of D-to-A converters is available on small computer plug-in boards. Before buying a new one you must specify several parameters to get a DAC that is suitable to your specific applications. The various selection criteria and the range of options available in commercial equipment are discussed below.

Range: Most DACs for computer interfacing applications have ranges of the order of ten volts. Typical ranges are 0 to 5, 0 to 10, -5 to 5, and -10 to 10 volts. Some boards allow the user to select the range under software control while others require changing the position of a jumper on the board. In applications requiring a different voltage range, say 0 to 0.1 volts, either an external amplifier or a multiplying DAC may be used. A multiplying DAC provides an external reference voltage input. Typically, the user supplies a voltage equal to the desired full scale output. A DAC may not be appropriate for applications requiring large voltage ranges, for example, 0 to 100 volts. In those cases it may be more convenient to use a programmable power supply.

Resolution: The resolution is defined as the size of the output voltage increment for a one-bit change in the input digital number. It is dependent on the total voltage range and the number of bits. The voltage increment is called one least significant bit or one LSB and is just equal to the total range divided by 2^n where N is the number of bits. The most common DACs available for general purpose use are either 8- or 12-bit, giving 256 or

40 Digital-to-Analog Conversion

4,096 output states respectively. Special purpose DACs with 6, 10, or 16 bits of resolution are available. The example in Figure 7.10 shows that the resolution is improved by a factor of 16 in going from an 8- to a 12-bit DAC. The example is for a converter that has an output range of -5 to +5 volts.

8 bit	**12 bit**
$1 \text{ LSB} = \dfrac{10 \text{ V}}{256} = 39.1 \text{ mV}$	$1 \text{ LSB} = \dfrac{10 \text{ V}}{4096} = 2.44 \text{ mV}$

Fig. 7.10 The measurement resolution of 8- and 12-bit DACs for -5 to +5 volt range

Speed: There are three different parameters relating to the speed of the DAC. The most commonly quoted is the data conversion rate, or the rate that at which are accepted from the computer bus and converted to an analog voltage. Another measure of the speed is the settling time, the amount of time it takes the converter's output to reach a stable value. You may be surprised to find that the settling time is often slower than the data conversion rate in high-speed DACs. Finally, in some high-speed applications you must consider the output slew rate, which describes the maximum time derivative of the output voltage. The output slew rate may limit the performance of high-speed DACs in applications where the input digital number changes rapidly from a small to a large number or vice versa. DACs may be purchased on plug-in cards for small computers with data conversion rates ranging from hundreds per second up to about 1 million per second. Even higher data conversion rates are possible, and current source DACs may be purchased with conversion rates in excess of 100 million/sec. However, these require custom hardware implementation and would not be convenient for general lab use.

DMA Capability: Direct Memory Access (DMA) is really a feature of the board holding the D-to-A converter, rather than the DAC itself. DMA means that data are transferred directly from the memory to the DAC rather than going through the processor. DMA is essential for high-speed operation where the two-step process of transferring data from memory to the processor and then from the processor to the DAC would be too slow. Autonomous operation of the DAC is also possible with DMA. This means that the processor can be performing some other function while the DAC is doing a series of conversions. This is particularly important for continuous data transfers where the DAC continuously cycles through an array of numbers. Continuous data transfer is really a software topic; to learn more about such transfers, read Section 7.4, DAC Software Overview.

Number of Channels: Analog inputs can be arranged so that a single A-to-D converter can be used to sample several different voltage inputs using a multiplexer to switch from one input to another. This same arrangement is more difficult to implement with analog outputs, so normally a single D-to-A converter drives only a single output channel. Thus, you need a separate DAC for each control voltage signal you want to send out. Multifunction I/O boards often come with more than one DAC; two seems to be a common number. Interface boards can be purchased with larger numbers of DACs for applications requiring many voltage outputs.

Current Sourcing Capacity: A normal DAC can supply just a few milliamps and still maintain the correct output voltage. Connecting any substantial load reduces the output voltage and may damage the DAC. Figure 7.11a shows a typical 8-ohm stereo speaker connected to the DAC output. DAC attempts to drive a 1-volt sinusoidal waveform into 8Ω load but cannot provide sufficient current. The result is a small sound. Figure 7.11b is identical to Figure 7.11a, with the addition of an amplifier to correct the problem. The

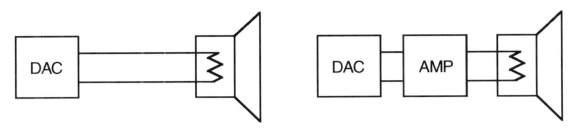

Fig. 7.11a Loudspeaker connected to DAC output   **Fig. 7.11b** Amplifier added to the system

amplifier can supply the necessary current (1/8 amp) so that the sound is louder. The actual sound difference can be heard on the HyperCard version of this book.

Timing Sources: A timing source is needed for multiple conversions. Different types of timing sources are available depending on the level of sophistication of the board you purchase.

Processor-Controlled: The DAC performs a conversion whenever the processor sends it a new number to convert. It is difficult to control precisely the rate of conversions with this technique.

Internal Timing Source: The board includes a programmable timing source so that you can request that conversions occur at any requested interval within the limits of the board's performance.

External Timing Input: There is a separate input pin allowing the user to supply a TTL compatible timing source.

7.4 DAC Software Overview

Various software routines must be used to control the D-to-A converter depending on the type of operation you want to perform. Specifically, the software must tell the DAC where to get its data, when to initiate conversions, and what the timing source will be. Five basic types of routines are required for a wide range of applications. They are described below. Note that not all types of routines are available for all hardware or software that you purchase.

Single-Sample Mode: Single sample is the standard operation mode for most DACs. On processor command a number delivered from the processor is converted to a voltage. The DAC holds that output voltage until the processor commands it to convert another number or the board is cleared. A wind tunnel speed control example is shown in Figure 7.12. A control voltage is sent to the motor controller via the DAC. The DAC holds that voltage with no attention from the processor, thus maintaining a constant motor speed. Whenever a new wind tunnel speed is desired, a different voltage value is sent out, and that value will be held until the next change.

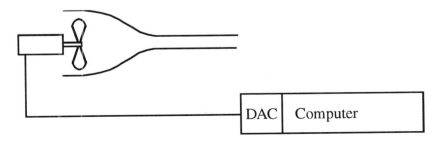

Fig. 7.12 Wind tunnel speed control as an example of single-sample conversions

42 Digital-to-Analog Conversion

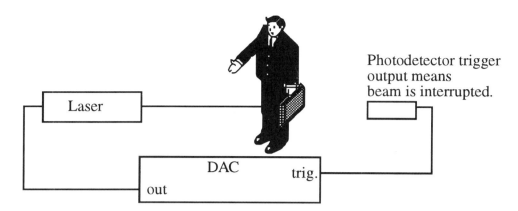

Fig. 7.13 Example of single-sample triggered operation. The DAC output voltage is 5 volts, but it is preloaded to send out 0 volts on receipt of a trigger signal from the photodetector.

Single-Sample Triggered Mode: The single-sample triggered mode is the same as the single-sample; except that the DAC is set up by the processor but doesn't actually perform the conversion until an external trigger is sensed. Once the DAC has been set up in this mode, it requires no further attention from the processor. The DAC will hold the previous output voltage until the trigger signal is received, at which time the DAC output will change almost immediately to the new value. The processor can check the board's Control and Status Register to determine if the conversion has occurred, but it does not have to do this.

The example shown in Figure 7.13 illustrates a typical application where the DAC is used to issue a response to an external stimulus. The DAC is preloaded with the correct value for the response. On receipt of the stimulus (trigger signal), the conversion occurs almost immediately. The example shows a laser whose output power is controlled by a voltage signal from a DAC. The light sensor will send a trigger pulse if the beam is interrupted. The DAC is outputting a high voltage that corresponds to a high laser power. However, it is preloaded to output a low voltage corresponding to low laser power upon receiving a trigger signal. The visitor enters the room, accidentally blocking the laser beam. The trigger signal occurs, and the lower voltage appears on the output in a few microseconds. The laser is thus shut down before it can harm the visitor.

Sequential Conversion: Sequential conversions means that an array of numbers stored in the memory are sequentially converted to voltages on successive pulses from the timing source. Each value in the array is converted only once. The timing source may be either internal or external. Normally, the timing source produces pulses at regular time intervals. For example, the timing source may be programmed to provide a pulse every 1/1,000th of a second so that the voltage changes occur precisely 1,000 times per second. This is particularly useful for motion control systems and other devices requiring a varying voltage input. The desired output waveform is approximated as a series or discrete samples. The array of numbers is then set up in the computer's memory and the DAC commanded to convert them sequentially. Figure 7.14 illustrates the DAC and the memory array. A memory pointer stored in the processor indicates the current sample being sent out.

Continuous Data Transfer: Continuous data transfers are the same as sequential conversions in that the elements of an array are sequentially converted. The difference is that the DAC loops through the array continuously. This is an excellent way to produce periodic waveforms. For example, the array stored in memory could define a single cycle of a sine wave. Running through this array repeatedly would imitate the function of a signal generator, producing a regular train of sine waves. Many DAC subsystems will run

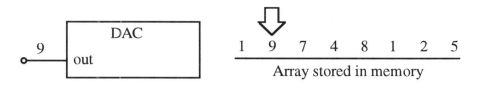

Fig. 7.14 Sequential conversion of an array of voltage samples

completely autonomously in this mode, meaning that the DAC can send out a periodic waveform indefinitely with no attention from the processor.

Double-Buffered Data Transfer: Double-buffering is a way to send out an array of values that is too long to store in the memory but must be stored on disk. The first chunk of data, say 2,000 numbers, is placed in memory and a continuous data transfer loop initiated. When the DAC completes 1,000 conversions the program refills the first half of the memory array with new values from the disk. Then, when the DAC completes the second 1,000 conversions, the program refills the second half of the array. This process continues until the entire array stored on disk has been sent out. The HyperCard version of this book shows an animation of this process, which helps clarify the explanation. The speed of double-buffered conversions is limited by the disk speed. For example, the disk must be able to refill the first half of the array before the DAC completes conversion of the second half.

7.5 LabView Control of the DAC

The LabView routines for operating the DAC are divided into three sets: Easy I/O VIs, Intermediate VIs, and Advanced VIs. The Easy I/O VIs have controls that are relatively easy to understand so that you can get started quickly in using the DAC. They may be used as either stand-alone programs or as subVIs in other programs. Intermediate VIs are more complicated to use but offer more flexibility in how you use the DAC. The Advanced VIs are quite complicated to use but allow you to have full control of all of the DACs features.

LabView is very limited in controlling the DAC when only an NB-MIO-16 board is present. Only the single-sample software mode described above can be used. Other software modes, such as sequential conversions and double-buffered data transfers, are supported using the LabView Waveform Generation VIs. However, these VIs require that an NB-DMA-8-G Direct Memory Access board be installed and linked to the multifunction board via the RTSI bus. This increases the cost of the system substantially.

The VIs described below make it very easy to produce analog output voltages after just a few minutes' practice. The VIs may be incorporated into larger virtual instruments, allowing you to include DAC output in your data acquisition and control programs. The following sections describe the LabView VIs to control the DAC and then lead you through a few exercises, first using the Easy I/O VIs as stand-alone programs and then incorporating them into more complicated programs.

There are four Easy I/O VIs to control the DAC as described in the LabView Data Acquisition VI Reference Manual. Only the first two described actually work with the NB-MIO-16 board we use in this tutorial.

AO Update Channel sends a voltage value to one DAC channel. The DAC holds the output at the specified voltage until it is updated again.

AO Update Channels sends voltage values to two or more DAC channels. Operation is otherwise identical to AO Update Channel.

AO Generate Waveform performs sequential conversions of a series of voltage values at a specified conversion rate.

44 Digital-to-Analog Conversion

AO Generate Waveforms is the same as AO Generate Waveform except that it is used for multiple channels.

Exercise 1: Using AO Update Channel

Your first use of the data acquisition VIs will be to send a voltage value to one DAC channel. First connect a voltmeter to DAC channel 0 so that you can confirm that you are sending out the correct voltage. Now you want to start up LabView and open the AO Update Channel VI. The easiest way to do this is to first open the LabView 3 folder, then open the VI.lib folder. Within the VI.lib folder, there are a number of VI folders and libraries. The libraries contain compressed versions of the VIs. Open the folder labeled DAQ. All of the Easy I/O VIs are in the 1easyio.llb library. Double-click on 1easyio.llb, and LabView will start. You will get a dialog box with a list of the Easy I/O VIs. You can choose to open any of the listed VIs at this point. Select AO Update Channel. You will now see the front panel and you can send voltage values to the DAC. Follow the directions below to set the controls and run the VI. Set a voltage of a few volts. The output voltage should be quite close to the voltage you specify. If your DAC is set up for a range of +/- 10 V, then the smallest voltage step is 4.88 mV. Try changing the specified voltage in steps of 0.002 volts, and observe the output on your voltmeter.

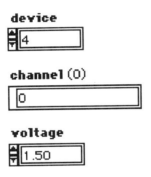

Fig. 7.15 Front panel for AO Update Channel

The front panel for AO Update Channel has the three controls shown in Figure 7.15. Device is the slot number where your data acquisition board is installed. If you don't know the slot number, pull down the menu below the apple at the top left of your screen and select Control Panels. Double-click on the NI-DAQ Utilities icon. This will display a chart showing where your data acquisition boards are installed. Channel is the selected output channel. If you are using an NB-MIO-16, this will be either 0 or 1. The voltage is the value that you are sending out. The possible range depends on how your board was set up when it was installed. It could be 0 to 5 volts, 0 to 10, -5 to 5, or -10 to 10. Go ahead and experiment; you can't hurt the board by specifying an incorrect voltage. In order to change the specified voltage in 0.001 volt steps, you will have to change the precision of the voltage control. Change to the Edit mode and pop up a menu on the voltage control. Select the Format and Precision option, and change it to three digits after the decimal point.

Use of the AO Update Channels VI

AO Update Channels is a little more complicated to use. The front panel controls are shown in Figure 7.16. Now you must specify a set of output channels. You do this by listing the channels separating the numbers by commas. You can also list the channels by separating the first and last channels of a sequence with a colon. For example, channels 0, 1, 2, and 3 can be listed 0:3.

LabView Control of the DAC 45

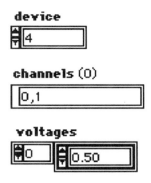

Fig. 7.16 *Front panel for AO Update Channels*

The voltage control is now a one-dimensional array. The control has two parts; the left number is the array index, and the right number is the voltage. The voltage value in the first element of the array is sent to the first channel listed, and so on. Test this with your voltmeter, and make sure you understand how to use the controls.

Using Easy I/O in a Diagram
You can use any of the Easy I/O VIs as part of a VI that you are building. When you are working on the block diagram, pop up the Function menu and select DAQ. Within the DAQ, select Easy I/O. The Easy I/O's palette will appear as shown in Figure 7.17. As you slide the arrow over each icon, the name of the icon appears at the bottom of the box. When you select one, the icon will appear on your block diagram. Select Show Help Window under the Windows menu. Move the arrow onto the icon, and the help window will show you a description of the VI and the icon with its terminals. The wires indicate the appropriate data type. The icon for AO Update Channel.vi and its help window are shown in Figure 7.18.

Exercise 2: Using Easy I/O in a Diagram
You now can use AO-Update Channel in your own more complex VI to practice basic LabView techniques and to learn how to incorporate the Easy I/O VIs into your own programs. The problem is to make a program that will slowly ramp a voltage up to a specified level. We might use such a program to smoothly accelerate a computer-controlled motor. Your VI should start with an output voltage of zero, then increase the voltage in steps of 0.005 volts until the selected voltage is reached. You will need to use A0 Update Channel repeatedly to change the output voltage in steps. It is probably easiest to do this with a While Loop. You should add a wait function to slow down the loop so that the ramp doesn't go too quickly. The Wait Until Next ms Multiple function under the Dialog & Date/Time menu is perfect for this.

The exercise is worked in Figure 7.19. Try to work it first without peeking. If you get stuck, go ahead and look at our solution. There are many different possible ways to program this, so don't worry if yours is different. Once you've got your VI working, observe the output signal on an oscilloscope.

The front panel includes controls for the device, channel number, and final voltage. The diagram is a while loop that continues to execute as long as the output voltage is less than the final voltage. The value of i is incremented each time the loop executes so that the voltage sent out increases in steps of 0.005 volts. The metronome icon is the Wait function used to slow the execution of the loop. With the input value of 100 it produces a wait of 100 milliseconds for each iteration of the loop. The constant value of 100 could be replaced with a panel control to vary the ramp speed.

46 Digital-to-Analog Conversion

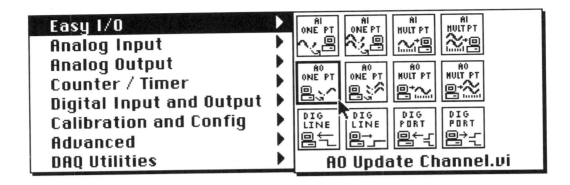

Fig. 7.17 The Easy I/O palette

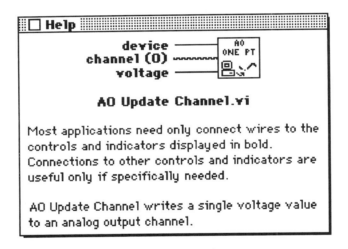

Fig. 7.18 The help window for AO Update Channel

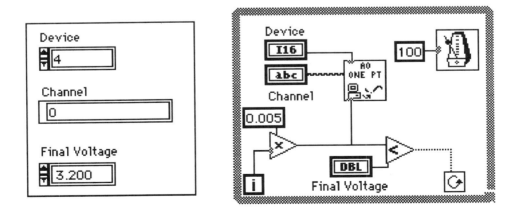

Fig. 7.19 The front panel and diagram for a solution to exercise 2

Intermediate and Advanced VIs

There are both intermediate and advanced sets of VIs to control the DAC. Most of these are used for sending arrays of data and cannot be used with the MIO-16 board. The three listed below are useful for building more complex VIs but they offer little advantage over the Easy I/O for sending out a single voltage. These few paragraphs offer only a brief introduction to the VIs. You'll have to read the manual to learn how to use all their features.

AO Write One Update is the intermediate VI that is used to send out a new voltage sample to each of the selected channels. You use it like AO Update Channels. The voltage and channel inputs must be arrays even if you are using only one channel. If you are using this VI in a loop, you should wire the iteration counter to the iteration terminal of AO Write One Update. If you do this, the DAC will be configured only once rather than on every iteration of the loop.

AO Group Config is an Advanced VI that is used to assign a group number to a list of channels. You need to use this before using any of the other Advanced VIs, even if you want to use only a single channel. AO Group Config also assigns a task ID number that is used as an input to the other Advanced Analog Output VI's.

AO Single Update updates each of the channels that were chosen using AO Group Config. The only inputs you need to connect are the task ID and a voltage array that contains as many elements as the number of channels selected.

Exercise 3: A Sine Wave Generator

It is unfortunate that we can't use the waveform generation VIs because one of the most useful functions of a DAC is to produce a repetitive waveform. We can try to do this by sending out single samples one after another. There are several problems with this. First, the user can't control the interval between successive samples. This just depends on how fast the computer can execute the voltage output operation. Also, the time interval between each set of successive samples will not necessarily be the same. Finally, this operation is slow, so only low frequency signals are possible.

Let's try to make a sine wave generator despite all these problems. Use a While Loop and a front panel control to allow you to run the loop for as long as you would like. You should represent the sine wave by a set of eight equally spaced values. That is, at each iteration the output voltage should be:

$$V_{out} = \text{amplitude} \times \sin(\pi i/4)$$

where i is the iteration number and amplitude is input via a front panel control. Look at the output signal on the oscilloscope to determine the frequency. The exercise is worked out in Figure 7.20 but don't peek until you've tried it.

The diagram is fairly simple. The two Build Array functions are needed because AO Write One Update requires arrays for the channel and voltage inputs. In this case, the arrays each have only a single element. Note that the iteration number i is connected to AO Write One Update so that the DAC is configured only once instead of every time the loop executes. The while loop continues to execute as long as the Run switch is turned on.

An alternate solution can be programmed using the Advanced VIs for the DAC. The alternate solution shown in Figure 7.21 executes about 20 percent faster than the other solution. The group configuration VI assigns the channel that will be used and produces a task id, which is needed by AO Single Update. The sine wave is computed in the same way as before. The single update function just updates the DAC output each time the loop executes. The channel is already preassigned. Try programming this VI, and compare the sine wave frequency to your earlier solution. You will find the two DAC VIs you need in the Advanced DAQ menu.

48 Digital-to-Analog Conversion

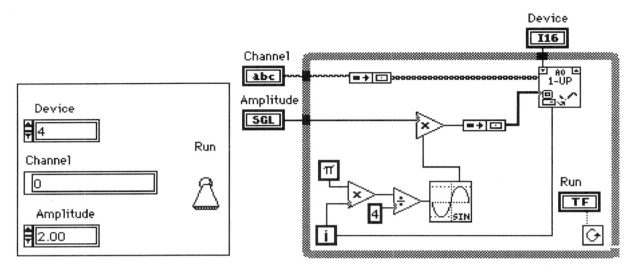

Fig. 7.20 Front panel and diagram for a solution to Exercise 3

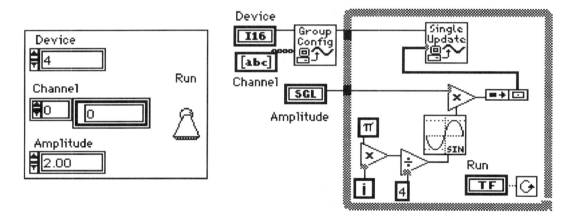

Fig. 7.21 Front panel and diagram for an alternative solution using Advanced VIs

7.6 Hardware Exercises

You may be getting a little bored just sending voltages out to a voltmeter or oscilloscope. Here is a system you can build quickly that will let you control the speed of a DC gearmotor. A gearmotor is a small, high-speed electric motor linked directly to a gearbox. Therefore, the output shaft turns fairly slowly but with considerable torque, even using a small motor. For example, we use a Dayton 2L009, which has a gear ratio of 394:1 in a probe-positioning system. Such a motor may be purchased for about $20. By varying the input voltage from 0 to 12V, you can vary the motor speed from 0 to 6700 revolutions per minute (RPM). The output shaft speed varies from 0 to 17 RPM. The maximum torque is 30 in-lbs even though the motor is only 1/125 horsepower. Similar motors may be purchased with a wide range of motor powers and output speed ranges.

The only problem with controlling DC motors with a DAC is that the motors draw a lot of current so cannot be powered directly by the DAC Output. For example, our motor can draw up to 1.4 Amps. You will need a special drive circuit illustrated in Figure 7.22.

LabView Control of the DAC 49

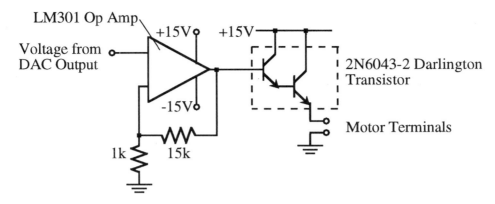

Fig. 7.22 Motor drive circuit

This circuit is controlled by the DAC output and supplies adequate current to power the motor. It consists of an operational amplifier with two resistors to control the gain and a Darlington transistor. These parts can be purchased at most electronic stores or from catalogs. You will also need a +/-15V power supply. Such a supply is very useful for many purposes in the lab. This circuit can drive loads up to 8 amps so you can use it to drive many different DC devices including lamps, fans, heaters, and a wide variety of motors.

Figure 7.23 illustrates a device to measure the speed that a fish can swim.* The propeller is driven by a DC motor controlled by the DAC and forces water around the loop. The flow speed may be increased by increasing the motor speed. A light beam is aimed across the channel and falls on a light sensor. Interruption of the light beam indicates that the fish is being swept backwards and that the motor speed should be reduced. At this point in *LabTutor* you know how to control the propeller speed, and soon you will know how to read the signal from a light sensor and a water velocity sensor.

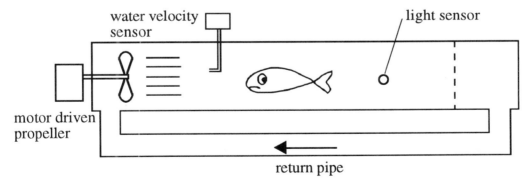

Fig. 7.23 An exercise machine for fish

---

* This was inspired by a visit to the lab of Professors I. Miaoulis and C. Rogers at Tufts University.

Chapter 8

Analog-to-Digital Conversion

8.1 Introduction to A-to-D Conversion

The function of an analog-to-digital converter is to convert analog voltages from the external environment to digital numbers that the computer's processor can use. It is much like a regular voltmeter in that it can be used to measure a voltage. A schematic diagram of a single-channel, n-bit A-to-D converter is shown in Figure 8.1. A voltage applied between the input terminal and the ground terminal is converted to an n-bit digital representation. The digital number is then passed over the computer's bus to the processor. The ADC may be used to read in a single voltage. Alternatively, the ADC may be used to read a series of voltage samples to record a time-varying voltage signal.

The A-to-D converter divides its input range into discrete steps. It measures the input voltage by locating the step that contains that voltage. The voltage measurement resolution is then equal to the size of the steps. The total number of available steps is determined by the number of output bits. For example, an 8-bit converter has $2^8 = 256$ possible output states and a 12-bit converter has $2^{12} = 4,096$ states. The larger the number of output bits, the more output states and the finer the measurement resolution. An 8-bit ADC with a 0 to 10 volt range measures the voltage with a resolution of 10V/256 = 39 mV. A 12-bit converter with the same 10V range has a resolution of 10V/4096 = 2.44 mV. Most ADCs used in computer interfacing applications have either 8 or 12 bits of resolution. Sixteen-bit ADCs are widely available, but they are more expensive.

The size of the discrete steps is called one least-significant bit, or one LSB. The ADC delivers its output to the computer in the form of an integer number that is equal to the number of LSBs the input voltage is above the minimum voltage. This integer number is easily converted to an actual voltage measurement in software.

A-to-D converters are used for reading voltages from a wide variety of transducers. Transducers are available to convert temperature, force, displacement, light level, angular position, surface velocity, gas velocity, rotational speed, pressure, or strain to a voltage signal. With appropriate signal conditioning (see Section 8.2 on hardware) the ADC can read very small voltages, resistance, and current. Thus, with the right transducer and an ADC, the computer is capable of obtaining information about nearly any variable in its physical environment.

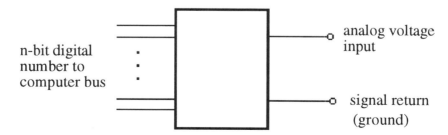

Fig. 8.1 Schematic representation of an analog-to-digital converter

A-to-D converters are also an internal component in many other systems. Digital multimeters use a high resolution ADC, typically around 24 bits. These ADCs are highly accurate but are slower than the 12-bit ADCs usually used in computers. Digital oscilloscopes use very high speed ADCs which typically have four- or six-bit resolution. In consumer electronics, A-to-D converters are used in automotive controllers, digital video recorders, and a wide variety of other applications.

The transfer function is the relationship between the input analog voltage and the output digital number. The integer number delivered to the computer must be converted to a voltage using the transfer function in order for the program to make sense of it. The transfer function is typically written:

$$\text{Output Number} = \frac{V_{in} - V_{min}}{V_{step}} \qquad V_{step} = \frac{V_{max} - V_{min}}{2^n}$$

where V_{max} and V_{min} are the maximum and minimum input voltages respectively and n is the number of bits. Figure 8.2 shows the transfer function for an ideal three-bit ADC with a 0 to 8V range. Note that the transition from output state 0 to output state 1 occurs at 1/2 volt (1/2 LSB). Therefore, the indicated voltage is always within 1/2 LSB of the input voltage.

8.2 ADC Hardware Overview

A key element of all A-to-D converters used in computer interfacing applications is the comparator. (See Figure 8.3.) The comparator is a simple circuit with two analog input terminals. It compares the voltage between its two terminals. The single digital output goes to the high state if the voltage is higher on the + terminal, and the output state is low if the voltage on the - terminal is higher.

There are two main types of A-to-D converters used in computer interfacing applications: combinational and sequential ADCs. A combinational ADC has one comparator for every possible output state and performs the conversion in a single step. A

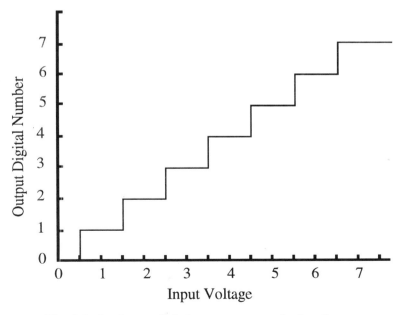

Fig. 8.2 Analog-to-digital converter transfer function

52 Analog-to-Digital Conversion

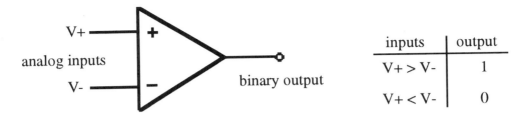

Fig. 8.3 Schematic representation of a voltage comparator

sequential ADC uses a single comparator in conjunction with a DAC and a logic circuit and requires several steps to determine the input voltage. Combinational ADCs are then characterized as low resolution devices (not very many bits) that are extremely fast. Conversion rates in excess of 100 million samples/sec are possible, and resolutions up to eight bits are available. Sequential ADCs are slower, and resolutions up to 16 bits are widely available. Twelve-bit sequential ADCs operating at speeds up to 1 million samples/sec are common in computer interfacing applications.

A third-type of converter, which has even higher resolution, is called a dual-slope ADC. These converters achieve the very high resolution essentially by integrating the input voltage over a short time. As a result, the dual-slope ADC is quite slow. These are used in digital voltmeters but not in computer interfacing applications.

A three-bit flash converter with a 0 to 8 V range is illustrated in Figure 8.4. This ADC will produce the transfer function illustrated in the introduction to this chapter. The flash converter is built using seven comparators and also requires seven voltage sources. In practice, the seven voltage sources are replaced with a single reference voltage source and a set of resistors, forming a voltage divider network. In the example shown, the input voltage is 4.72 volts, so comparators 1 through 5 have output state 1, while comparators 6 and 7 have output state 0. The encoder is a simple digital logic circuit that puts out a three-bit number representing the position of the highest comparator to output a 1. Digital logic circuits like the encoder can operate very rapidly, which accounts for the high speed of the flash converter. Higher-resolution flash converters are available. Six-bit converters using 63 comparators and 8-bit converters using 255 comparators are both widely used.

Until recently flash converters were used only in special applications such as digital video and radar. They are difficult to use in small computer applications because the computer bus cannot transfer the data from the ADC to the memory as fast as the ADC acquires it. Special-purpose circuit boards that include a dedicated high-speed memory are available, but they are very expensive. These are generally used only in special digital signal processing applications.

A sequential ADC uses a single comparator, a DAC, and a logic circuit to measure the input voltage. The most common type is the successive approximation ADC shown in Figure 8.5. The successive approximation register is a logic circuit that alters the output digital number, then watches the comparator output to determine if the resulting DAC output is above or below the input voltage. The successive approximation algorithm discussed below is the fastest converging sequential algorithm and is used on virtually all sequential ADCs.

A successive approximation ADC works by successively dividing the input range in two. The first step is to set the most significant bit of the output to one, then compare the input voltage to the DAC output. If the comparator indicates that the input voltage is higher than the DAC output, we know that the input is in the upper half of the voltage range and we leave the bit set to one. If the input voltage is in the lower half of the input range, we reset the most significant bit to zero. We then follow the same procedure setting the next most significant bit to one. This determines which quarter of the range the input is in.

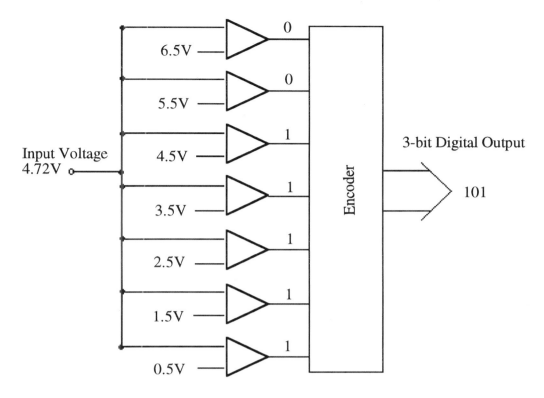

Fig. 8.4 A three bit flash converter with a 0 to 8 volt range

This process continues until we reach the least significant bit. Therefore, the successive approximation algorithm converges to the correct digital value in n steps, where n is the number of bits of resolution.

The chart in Figure 8.6 gives an example of the successive approximation algorithm in action. This example is for a four-bit ADC with a range of 0-16 volts (1 LSB = 1 volt) converting an input voltage of 5.92 volts. The chart shows the DAC output voltage as it changes during the conversion. After the first step, the DAC voltage is 8 volts. The comparator shows that this voltage is too high, so the most significant bit is reset to zero and the next bit is set to one. This produces a DAC voltage of 4 volts, which is less than the input voltage. The third bit is then set to one producing a voltage of 6 volts, which is higher than the input. The third bit is reset and the least significant bit set to one, producing a voltage of 5 volts. The comparator shows that this output voltage is less than the input, so the least significant bit remains set. The final digital output number is 0101, indicating a measured voltage of 5 volts.

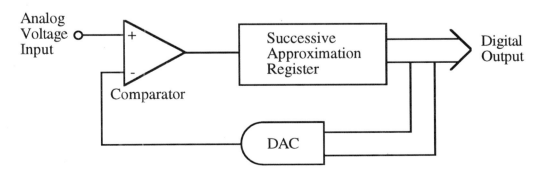

Fig. 8.5 Schematic of a successive approximation ADC

54 Analog-to-Digital Conversion

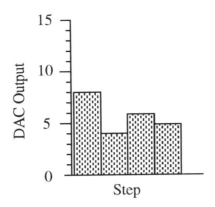

Fig. 8.6 Steps in the convergence of a successive approximation ADC to the correct measurement

8.3 ADC Auxiliary Hardware

An analog-to-digital conversion board that plugs into a small computer's bus requires auxiliary hardware besides the ADC itself. Generally the board will contain a single ADC connected via a multiplexer to multiple input channels. The board needs a bus interface unit and control logic to obtain data and control words from the computer's system bus. The bus interface includes an interrupt system and may also include the hardware to allow the ADC to transfer data via a direct memory access channel. In addition, the ADC needs a timing source to allow voltage measurements at precise time intervals. All of this is the same as for the DAC and is described in Chapter 3.

The ADC also requires some additional hardware including one or more sample and hold amplifiers to freeze the input signal while the conversion is being done, a multiplexer to connect the ADC to multiple input channels, a variable gain preamplifier to allow the ADC to measure accurately low-level voltage signals, and a FIFO buffer memory to temporarily store samples before they are transferred to the processor or memory.

Most A-to-D conversion subsystems for use with small computers come equipped with multiple input channels, typically 8 or 16. These subsystems usually contain only one ADC and use a multiplexer to gain multichannel capability. A multiplexer is just a computer controlled switch as illustrated in Figure 8.7. Plug-in boards allowing as many as 64 analog voltage inputs are available for PCs. External multiplexers allowing a huge number of inputs to a single ADC are also available.

It is essential that the input voltage remain constant during the process of analog-to-digital conversion. A sample-and-hold amplifier such as the one illustrated in Figure 8.8 is used to hold the voltage during the conversion. The output follows the input as long as the switch is closed. After the switch opens, the output remains nearly constant, decaying slowly towards zero. In operation the switch is normally closed. The switch opens when

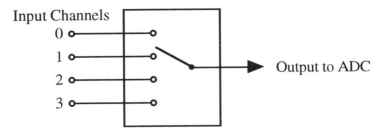

Fig. 8.7 A four-channel multiplexer

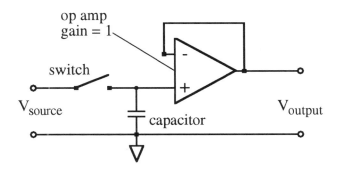

Fig. 8.8 Schematic of a sample-and-hold amplifier

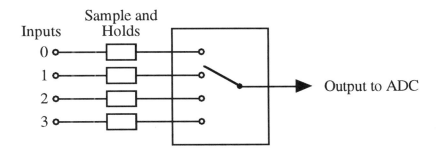

Fig. 8.9 A four-channel multiplexer with individual sample-and-holds

the timing source triggers a voltage sample. Most ADC subsystems you would purchase are equipped with a single sample and hold located between the multiplexer and the actual ADC.

In many applications it is necessary to sample two or more voltage inputs simultaneously. This cannot be done with an ADC and a multiplexer alone because the ADC can convert only one voltage at a time. The best ADC subsystems include a separate sample-and-hold for each input channel. The sample-and-holds are triggered simultaneously, then the inputs converted successively.

Data may be lost from a high-speed A-to-D if the computer's bus is not available to transfer data from the A-to-D to memory before the next data sample is converted. High-speed A-to-D boards usually include a small buffer memory to store a few data samples until the bus becomes available. The memory is called a First-In-First-Out, or FIFO, because the data are taken out in the same order they were put into the memory. A typical FIFO on an A-to-D board has about 16 storage cells.

If the voltage you would like to measure is small, even a 12-bit A-to-D converter may not have sufficient resolution to measure the voltage accurately. Most A-to-D subsystems include a preamplifier to alleviate this problem. The preamp gain is software programmable

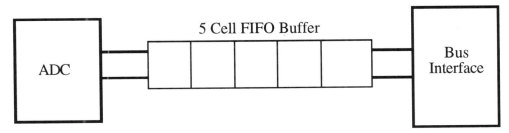

Fig. 8.10 A First-in-First-Out (FIFO) buffer for the ADC

56 Analog-to-Digital Conversion

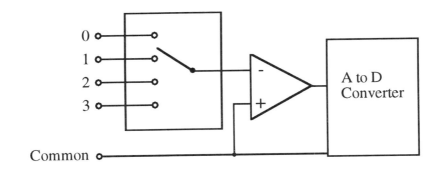

Fig. 8.11 A four-channel multiplexer with a preamplifier

on some boards, while others require the user to remove the board from the computer and change a jumper in order to change the gain.

8.4 Selection Criteria

A-to-D converters are available on interface boards from a large number of manufacturers. A wide range of features is available, and the price may vary widely. You must specify several parameters, such as resolution and conversion speed, before purchasing a new board for your computer. This section discusses the various factors you need to consider. You should also consider some of the optional equipment discussed in the Auxiliary Equipment section.

Range: A-to-D converters usually have ranges on the order of 10 volts and are typically set up with bipolar inputs. The word <u>bipolar</u> means that the ADC can accept either positive or negative voltages. A very common range is -10V to +10V but ranges of -5V to +5V and 0 to 10 volts are also common. A preamplifier is available with most ADC systems. The preamplifier amplifies the signal before conversion, so it decreases the range and increases the resolution of the converter. For example, consider a 12-bit ADC with a range of -10V to +10V, giving a resolution of 4.88 mV. Use of a preamp with a gain of 10 reduces the useful range of the converter to -1V to +1V but improves the resolution to 0.488mV. The preamp gain is software-selectable on some ADCs; it can be changed by moving jumpers on other ADC boards. If you are planning to measure small voltages such as from a thermocouple, you must make sure that the ADC you purchase has a high gain (say 500) setting available. An external preamplifier can be used if necessary, but this is usually an expensive and relatively inconvenient option.

Resolution: The resolution is defined as the size of the voltage increment for a single bit change in the output digital number. This voltage increment is called one least significant bit, or one LSB. The resolution is equal to the total voltage range of the ADC divided by 2^n, where n is the number of bits. Most general purpose ADCs have 12-bit resolution, thus dividing the input range into 4,096 steps.

Eight-bit ADCs are also common in computer interfacing applications, and 16-bit converters are available. Converters with even higher resolution are available but are used only for special-purpose applications such as digital voltmeters. The addition of a preamp increases the resolution and decreases the range, as illustrated in Figure 8.12.

<u>**Preamp Gain = 1**</u>

$$1 \text{ LSB} = \frac{20\text{V}}{4096} = 4.88 \text{ mV}$$

<u>**Preamp Gain = 8**</u>

$$1 \text{ LSB} = \frac{20\text{V}}{4096} * \frac{1}{8} = 0.61 \text{ mV}$$

Fig. 8.12 Change in resolution of a 12-bit ADC with -10V to +10V range using a preamp

Conversion Rate: There is a wide range of different conversion rates available in ADC subsystems for small-computer usage. Interface boards using successive approximation converters are available with conversion rates ranging from a few hundred samples per second up to about 1 million samples per second. Boards using flash converters are not widely available and are very expensive but provide much higher conversion rates. There are two precautions in reading conversion rate specifications. High-speed ADCs may require you to purchase an optional DMA controller in order to achieve the highest quoted conversion rates. Also, multichannel operation may slow the operation of some systems considerably. Read the entire specification sheet carefully before purchasing an ADC for a high-speed operation.

DMA Capability: Direct Memory Access (DMA) is really a feature of the board holding the A-to-D converter, rather than the ADC itself. DMA means that data are transferred directly from the ADC to the memory, rather than going through the processor. DMA is essential for high-speed operation where the two-step process of transferring data from the ADC to the processor, then from the processor to memory would be too slow. Autonomous operation of the ADC is also possible with DMA. This means that the processor can be performing some other function while the ADC is doing a series of conversions.

Number of Channels: Most ADC subsystems contain only a single A-to-D with a multiplexer to increase the number of input channels. Systems with 8 or 16 input channels are very common. If you need a large number of input channels, choose a system that allows you to add additional multiplexers without adding another ADC. Some manufacturers' systems are expressly set up to access a very large number of channels using multiplexers mounted in a separate box. If you need to access many channels, make sure that the board and software you purchase allow you to access easily all the inputs you need. In some applications where you want simultaneous sampling on two or more input channels, the critical question to ask is "How many channels have independent sample and holds?" In some cases you can buy a separate board that includes a sample-and-`hold for each input channel. If you have a high-speed operation you may need to consider how multichannel operation degrades the speed. For example, sampling a sequence of channels may be slower than taking a series of samples from a single channel on some systems.

Differential vs. Single-Ended Inputs: The drawing in Figure 8.13 shows an ADC set up with single-ended inputs. The voltage inputs are connected between the numbered input pins and the shared common line. Single-ended inputs are useful if all inputs share a common ground at the source. However, electrical noise on the common line introduced by one source will affect all of the voltage readings. Differential inputs are preferred for reliable measurements. With true differential inputs two multiplexers are used one for the signal lines and one for the signal return lines. The amplifier is sensitive only to the difference in voltage between the two terminals, so any common-mode noise is rejected. The disadvantage is that you get one-half as many input channels with the same hardware.

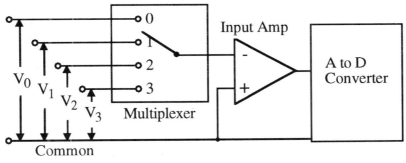

Fig. 8.13 An ADC set up with single-ended inputs

58 Analog-to-Digital Conversion

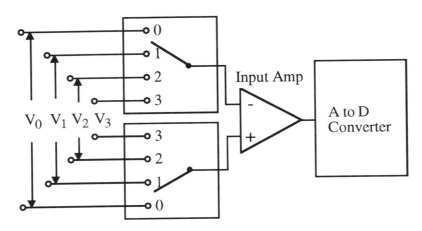

Fig. 8.14 An ADC set up with differential inputs

8.5 ADC Software Overview

The various types of software routines that are used to control the A-to-D converter are generally the same as those used for a D-to-A converter. The classes of routines are summarized below. For a more detailed description of any of these routines, look under the same headings in Chapter 7.

Single-Sample: ADC collects a single voltage sample on command of processor.

Single-Sample Triggered: ADC collects a single voltage sample upon receipt of an external trigger.

Sequential Conversion: ADC collects a series of voltage samples at regular time intervals and stores the samples in a memory array.

Continuous Data Transfers: Same as sequential conversion except that the memory array is repeatedly refilled. This can be done with an ADC but is not generally a very useful routine.

Double-Buffered Transfers: A memory array is continuously filled by the ADC. When the array is half filled, the data are transferred to disk storage. This allows the acquisition of a very long string of data.

The only new software problem in A-to-D conversion relative to D-to-A conversion is the specification of a channel sequence. The user may wish to sample repeatedly a set of channels rather than just one channel. Most software routines allow the user to specify the order of sampling.

8.6 LabView Control of the ADC

The LabView virtual instruments for operating the ADC are divided into three categories: Easy I/O, Intermediate VIs and Advanced VIs. Easy I/O VIs are simple to use but do not allow control over all parameters. Intermediate and Advanced VIs allow more flexibility but can be difficult to use. The VI's are similar to those available for the DAC but in this case you will be able to use almost all of the VIs with the NB-MIO-16 board.

The Easy I/O VIs are all you will need for most applications. We will go through several examples that use these VIs directly and incorporate them into more complex VIs. Most of the examples are actually worked exercises. You should do each of these exercises as you go through to ensure that you fully understand the controls.

Programming using the intermediate and advanced VIs can be a lot more complicated. As you work through the exercises you may also want to use the LabView Data Acquisition VI Reference Manual. Another way to understand the use of the advanced VIs is to open the diagram of an Easy I/O VI. Each of them is written using the intermediate and advanced VIs.

There are four Easy I/O VIs to control the ADC as described in the LabView Data Acquisition VI Reference Manual.

AI Sample Channel: Acquires a voltage sample from a specified channel and displays it on the front panel.

AI Sample Channels: Acquires single voltage values from two or more channels and saves the results as an array. The operation is otherwise identical to AI Sample Channel.

AI Acquire Waveform: Performs timed sequential conversions to acquire a series of voltage samples from a single channel and stores the results in an array. The sampling rate is controlled by the user.

AI Acquire Waveforms: The same as AI Acquire Waveform except this is used for multiple channels. The results are stored as a 2D array.

The AI Sample Channel VI

The front panel for AI Sample Channel has four controls as shown in Figure 8.15. The device is the slot number where your data acquisition board is installed. If you don't know the slot number, pull down the menu below the apple at the top left of your screen and select Control Panels. Double click on the NI-DAQ Utilities icon. This will display a chart showing where your data acquisition boards are installed. Channel is the selected input channel. The high and low limits are the maximum and minimum voltage values you plan to acquire. The VI determines the gain values using these limits. The gain can be 1, 2, 4, or 8 for a NB-MIO-16H or 1, 10, 100, or 500 for an NB-MIO-16L. If you have an NB-MIO-16L with the jumper set to a range of -10V to 10V and you set your high/low limit to +/-10V, your gain value will be 1. If you change your high/low limit to +/-1.0V, your gain value will be 10. The measured voltage is displayed on the front panel.

The icon and its connectors for AI Sample Channel are shown in Figure 8.15. The wires indicate the appropriate data type for each input. The channel input requires a string, and the other inputs require a number. The values in parentheses are the default values that are used if the terminal is not connected. To make a measurement on channel 0 with limits of -10 to +10 volts, you need only wire the device input.

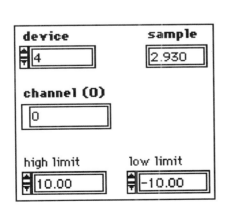
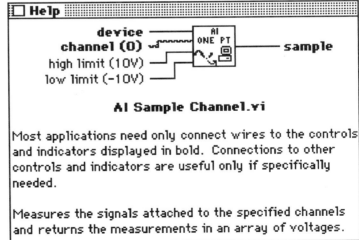

Fig. 8.15 The front panel for AI Sample Channel and its help window

60 Analog-to-Digital Conversion

Exercise 1: Reading a Voltage

The purpose of this exercise is to practice using the AI sample channel VI and to set up a system needed to test the more advanced VIs. You need to use a power supply or other stable DC voltage source (a battery will work fine) and a voltmeter so that you can supply a known voltage to the ADC. Set them up as sketched in Figure 8.16. Turn the power supply on, and set a voltage of about 3 volts. Measure the voltage using the voltmeter before connecting to the ADC. You can blow out the ADC by supplying a voltage that is beyond the input range of +/- 10 V. Now measure the voltage using AI Sample Channel.

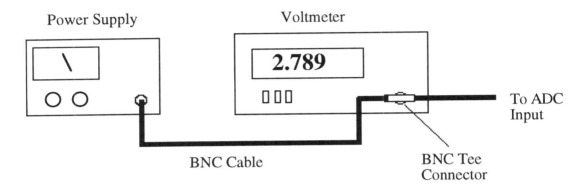

Fig. 8.16 Lab set up for Exercise 1

Exercise 2: Reading a Low-Voltage Signal

We next want to generate a low-voltage signal in order to understand the use of the limits in the AI Sample Channel VI. The easiest way to do this is to make a voltage divider circuit using two resistors as sketched Figure 8.17. The exact values of the resistors aren't important, but don't make the values too small or too much current will flow. Using the circuit shown, the measured voltage will be approximately 1/100th of the power supply voltage. Set the high limit on AI Sample to 10 V and the low limit to -10V. Change the format and precision of the voltage indicator to display 5 digits.

Now set the power supply to give a very low voltage so that the ADC input voltage is within 1 millivolt of zero. You should get an output on the panel indicator of 0.00000. Slowly increase the power supply voltage while repeatedly running the VI. Eventually the output indicator should switch to 0.00488. This shows that the resolution of the ADC is

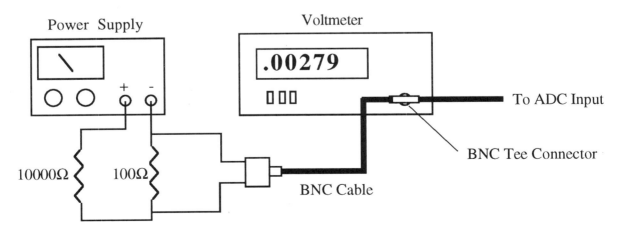

Fig. 8.17 Lab set up for Exercise 2

4.88 millivolts when a gain of 1 is used.

Next, increase the gain on your preamp. You do this in AI Sample Channel by changing the high and low limits. Set the high and low limits to + and - 1 volt. If you have an NB-MIO-16H you will get a gain of 8 and if you have an NB-MIO-16L you will get a gain of 10. Once again, slowly increase the input voltage from zero while running AI Sample Channel repeatedly. You will see how the resolution improves when you increase the gain. Note, though, that you have reduced the usable range of the ADC.

Exercise 3: An Autoranging Voltmeter

One of the advantages of using a digital voltmeter is that it can be operated in an automatic ranging mode. The output scale is automatically adjusted to account for the signal level. If a high-voltage signal is input, then a high-voltage range display is used, and vice versa. The disadvantage of the voltmeter is that it takes a long time (compared to an ADC) to make the voltage measurement.

We can do the same thing with the computer and an ADC. Write a VI that will measure the input voltage using the highest voltage range available. The VI should then select the maximum possible gain and re-measure the voltage. It will probably be easiest to do this using a case structure, with each case calling AI Sample Channel with a different input range. The output should be the final measurement of the voltage.

This exercise will be a good chance to practice your LabView programming skills, so don't peek at the answer unless you are stuck.

Our solution is a little bit complicated. This VI assumes that the available gains are 1, 10, 100, and 500. The voltage is first measured using the default range of -10 to +10 volts. If the absolute value of the voltage is less than 1 volt then the voltage is re-measured with a new range of -1 to + 1 volts. If the absolute value of the new voltage is less than 0.1 volts, then the voltage is re-measured with a range of -0.1 to 0.1 volts. Finally, if the absolute range of this voltage is less than 0.02 volts, the voltage is re-measured with the smallest possible range.

The diagram in Figure 8.18 shows only the true cases, that is, those in which the voltage is smaller than the comparison value and must be re-measured. In case the voltage

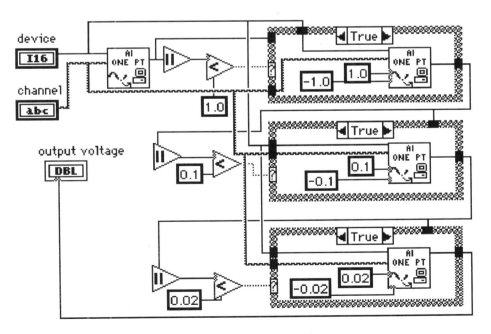

Fig. 8.18 Exercise 3: Block diagram

62 Analog-to-Digital Conversion

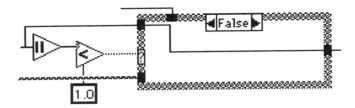

Fig. 8.19 Exercise 3: False case structure

is above the limit, we cannot use the higher gain and the present voltage value is just passed through the case structure as shown in Figure 8.19.

The AI Acquire Waveforms VI

The front panel and the icon for AI Acquire Waveforms are shown in Figures 8.20 and 8.21 respectively. This VI acquires multiple samples from the channels you select at the rate specified in scan rate. The channel list is specified by a string, with the channel numbers separated by commas. You can also list the channels by separating the first and last channels of a sequence with a colon. For example, channels 0, 1, 2, and 3 can be listed 0:3. The number of samples and the sample rate must be specified. The sample rate is the rate at which each channel is sampled. If you have four channels selected and a scan rate of 1000/sec, the ADC will actually be performing 4,000 conversions per second. The

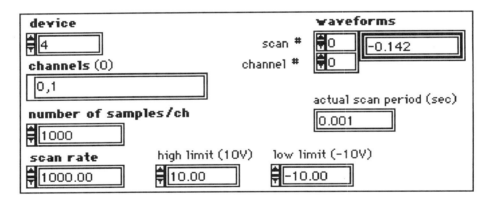

Fig. 8.20 The front panel for the AI Acquire Waveforms VI

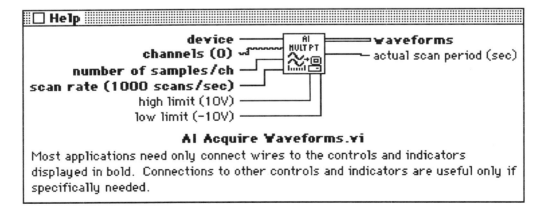

Fig. 8.21 The help window for the AI Acquire Waveforms VI

actual sample rate may not be exactly the same as the sample rate requested, so one output of the VI is the actual sample rate per channel.

The controls for device and high/low limits are specified as in AI Sample Channel. The output is displayed in an array form. To read the array from the front panel you must use the index controls to the left of the array value. Usually you will connect the output of this VI to another VI to further process or plot the data.

Exercise 4: VI to Acquire and Plot Voltages

We now will make practical use of the waveform acquisition VIs to simulate an oscilloscope, using the computer. In many cases it is actually more convenient to sample and display a waveform using the computer than to do so on the oscilloscope. Use the AI Acquire Waveforms VI and the graphing functions to make the equivalent of a two-channel oscilloscope. The two channels should be sampled at a rate that is selectable on the front panel. You can set the voltage limits just as you would adjust the volts/division control on your scope. Two lines should be displayed on the graph, one representing each of the two channels

You may have to review Chapters 3 and 4 in the LabView Tutorial. You will need to use the Bundle function to build a cluster and a waveform chart to display the data. Your VI should be fairly simple, so if it starts to get too complicated, go ahead and peek at our diagram.

Our VI, called VseqP, is shown in Figures 8.22 and 8.23. It acquires a sequence of voltage samples from specified channels. Its output is displayed in a two dimensional array and is plotted on a Waveform Graph on the front panel. The panel is similar to AI Acquire Waveform, and controls are specified in the same manner. The output array from the AI Acquire Waveforms function has the data for each channel stored in columns. However, the Waveform Graph expects the data for each channel to be in rows. The Transpose Array function is selected from the pop-up menu of the graph to solve this problem. Alternatively, you could transpose the array in the diagram by using the Transpose 2D Array from the Array & Cluster menu item of the Function menu.

The example shown in Figure 8.22 has only two channels. It is easy to acquire and plot more channels. The legend can be enlarged to show more plot labels, and the point styles, line styles, and colors can be selected using the pop-up menu on the legend.

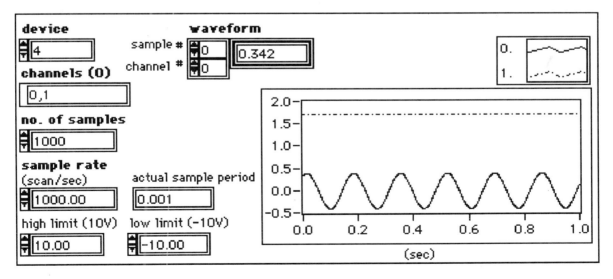

Fig. 8.22 Exercise 4: Front panel for the VseqP VI

64 Analog-to-Digital Conversion

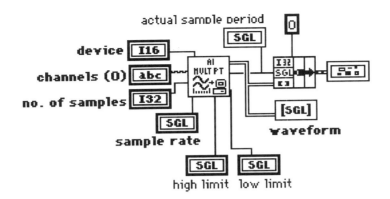

Fig. 8.23 Exercise 4: Block diagram for the VseqP VI

The diagram is really quite simple. AI Acquire Waveforms is connected to its inputs just as before. The voltage output is connected to the front panel array indicator and to a Bundle function. The Bundle function builds a cluster. This is needed because the Waveform graph requires three types of information: an initial x value, a Δx value, and the array of voltage samples. In this case x corresponds to time. The initial x value is 0, and the Δx value is the actual sample interval.

Set up the function generator to output a 200 hertz sine wave with a 1-volt peak to peak amplitude using the oscilloscope. Set the DC Offset to 0 volts. Now use VseqP to acquire and plot 100 samples at 10,000 samples per second. You should see two complete cycles of the waveform. Change the sampling rate and see the difference in the output graph. You can compare your results to those shown in Figure 8.24.

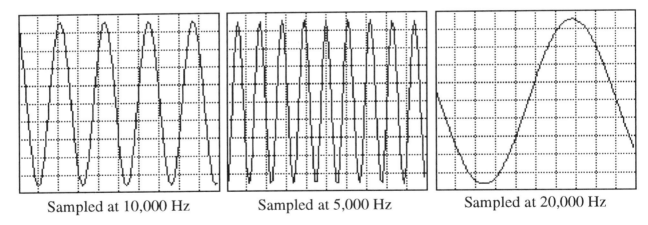

Fig. 8.24 Typical waveforms displayed by the VseqP VI

Understanding Multichannel Sampling

When you scan more than one channel, the scan rate you specify is the rate at which the board acquires samples for each channel. For example, in exercise 4, we used VseqP to scan two channels at 1,000 scans per second. This means that each channel is sampled at a rate of 1,000/sec and the ADC must acquire a total of 2,000 samples per second. Ideally, we wish to sample all channels simultaneously for each scan, but this is impossible with most data acquisition boards. Unless your board has a sample-and-hold on each input, there will be an interchannel delay time between samples on successive channels, as illustrated in Figure 8.25. The diagram shows the exact times when samples are acquired

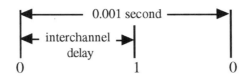

Fig. 8.25 Illustration of interchannel delay

on two channels, 0 and 1, which are being sampled with a scan rate of 1,000 Hz. The interchannel delay varies with the type of board you have and with the scan rate. You can specify the delay time using Intermediate Analog Input VIs, but this gets a little complicated. This will be covered in Chapter 11.

We next shown how to measure the interchannel delay time using VseqP by putting the same signal into two different channels. Set up a function generator to supply a triangle wave at 50 Hz and use a tee-connector to hook up the output to two different channels. (See Figure 8.26.) Set the scan rate of VseqP to 1,000 scans per seconds. The front panel for this setting is shown in Figure 8.27. The scale on the graph has been adjusted to show the first several samples clearly. If both of the channels were acquired simultaneously, the two triangle waves would overlap perfectly. Looking at the graph, we can see that if we slide the wave of channel 1 about half the sample period to the right, the waves would match. This indicates that the interchannel delay is about half of the actual sample period, or about 0.0005 seconds. To calculate the delay, we need first to calculate the slope of the wave shown in the graph. This was done using samples 2 and 9 of channel 0. The values are -.394 and .319 volts for sample number 2 and 9 respectively. Sample numbers 0 and 1 were not used, since it is not known for certain that these data points are part of the upward slope of the triangle wave.

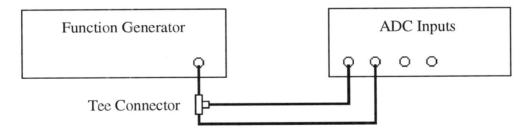

Fig. 8.26 Connections for testing interchannel delay

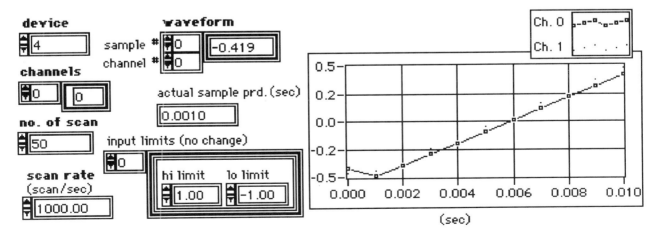

Fig. 8.27 Front panel of VseqP during measurement of interchannel delay

66 Analog-to-Digital Conversion

The slope calculation is:
slope = [0.319 - (-0.394)]/(7 X 0.001) = 102 volts/sec.
Using this slope value, we can calculate the interchannel delay time:
delay time = (data value for ch. 1 - data value for ch. 0)/slope.
The data values used in this equation must be from the same scan period. Using sample number 7 from both channels, the lag time turns out to be 0.00051 seconds, which is about half of the scan rate.

This exercise was repeated using a scan rate of 100 scans per second and a 5 Hz triangle wave. The interchannel delay came out to be 0.005 seconds, which is half of the sample period. This shows that the interchannel delay for the NB-MIO-16 board is half of the sample period.

Exercise 5: Mean and Standard Deviation

A very common use of the ADC is to measure the average value and the standard deviation of a voltage signal. For example, the force transducer on the model airplane engine test stand shown in Figure 8.28 will probably provide a fluctuating output signal because of electronic noise and actual variations in the thrust. The best way to measure both the mean and the standard deviation of the signal is acquire a set of voltage samples with the ADC and then calculate the statistics in the computer. Write a VI that uses the Easy I/O to acquire the voltages from a single channel and calculates the statistics using the Mean and Standard Deviation VI from the Analysis menu. Save this VI, because you inevitably will use it often in the laboratory.

The panel of our VI, called Vave is almost the same as in the previous exercise. The diagram is simple, consisting of just AI Acquire Waveform and the Mean and Standard Deviation function from the Analysis menu. (See Figure 8.29.)

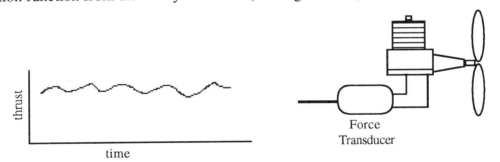

Fig. 8.28 A model airplane engine on a thrust test stand and a typical thrust vs. time trace

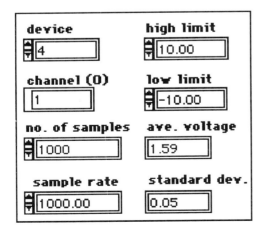
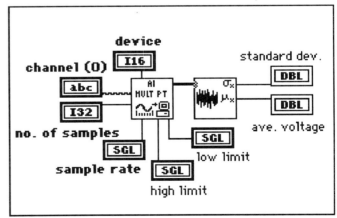

Fig. 8.29 Front panel and diagram for the Vave VI

Using Intermediate VIs

The Easy I/O VIs are quite simple to use and can be used for all the needs of most users. The intermediate and advanced VIs offer the user more flexibility in controlling the ADC during data acquisition, allowing such features as triggered operation and double-buffered data acquisition. Also, the intermediate and advanced VIs may run repetitive operations considerably faster than the equivalent Easy I/O. However, the more advanced VIs are more difficult to understand, and we suggest that you not study them in detail until you have become quite familiar with LabView programming and have a definite need for the more advanced capabilities.

The following descriptions and exercises give a short introduction to the intermediate VIs and will give you some familiarity with the programming. Exercise 6 may be skipped without any loss of understanding. However, exercises 7 and 8 should be worked using Easy I/O if you don't want to use the intermediate VIs.

If you are going to begin serious programming using the more advanced VIs you should first read the second half of Chapter 1 in the LabView Data Acquisition Reference Manual. You will then need to use the reference sections of the same manual (Chapters 4 and 9) to understand the individual VIs. The manual is difficult to understand without examples. Some examples may be found in the DAQ section of the LabView Examples folder.

AI Read One Scan: This is used to acquire one voltage measurement from each of the selected input channels. It is similar to the AI Sample Channels in the Easy I/O. You can choose to have the voltage output as a scaled voltage or as the binary value actually acquired by the ADC. This VI is better than the Easy I/O if used in a loop because it can be wired so that the ADC will be configured only once instead of each time through the loop.

AI Waveform Scan: This is the same as AI Acquire Waveforms in the Easy I/O, with two main advantages. It can be configured so that the ADC is configured only once if you are using it in a loop, and it allows triggered operation. For example, the set of samples may be acquired after receiving a trigger signal on a digital input line.

AI Continuous Scan: This VI is used if you want to fill continuously a circular buffer with voltage samples, usually for double-buffered transfer of samples to a disk file. You use AI Continuous Scan in a loop. The first call starts the data acquisition and returns a set of samples. Each subsequent call returns the next set of samples. For example, if you want to acquire a very long series of voltage samples at a uniform sampling rate, you use AI Continuous Scan. The ADC runs continuously, so your samples are always uniformly spaced in time. Each time you call the VI it returns the next set of samples.

AI Config, AI Start, and AI Read: These three VIs can be combined to do a data acquisition operation where the samples are stored in a memory buffer and then read out after the sampling is complete. AI Config sets up the operation, AI Start is used to trigger the start of the sampling, and AI Read is used to read the samples from the computer's memory once the sampling is complete.

Exercise 6: Using Intermediate VIs to Acquire and Plot Voltages

In this exercise you will reprogram exercise 4 using the intermediate VIs instead of Easy I/O. The point is to learn how to follow the manual in selecting and connecting an appropriate intermediate VI. You will need to use the Data Acquisition VI Reference Manual. After you have completed the VI, you should time its operation. You will find that the new VI is just slightly faster. In this case, the Easy I/O is just as effective as the intermediate, so there really is no reason to use the intermediate. However, you may need the intermediate VIs in the future when you make more complicated instrumentation systems.

Our solution, called VseqP2, uses an Intermediate VI called AI Waveform Scan instead of AI Acquire Waveforms. The running time for VseqP2 is slightly faster than that for VseqP, since AI Acquire Waveforms calls AI Waveform Scan. The front panel for VseqP2 is similar to that for VseqP. The channel input is an array of strings, and input limits is an

68 Analog-to-Digital Conversion

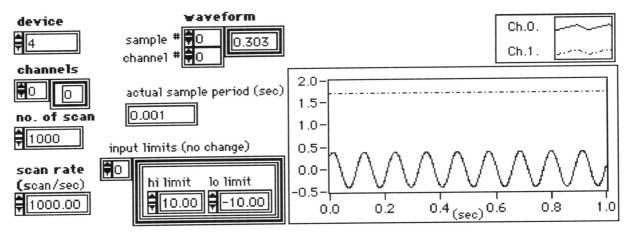

Fig. 8.30 Front Panel for VseqP2

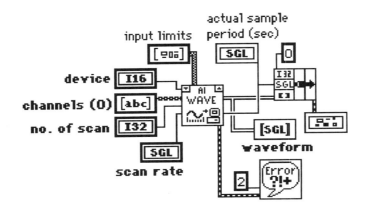

Fig. 8.31 Diagram for VseqP2

array of clusters. (See Figure 8.30.) The diagram for VseqP2 shows AI Waveform Scan connected to its inputs. It is very similar to the diagram for VseqP. (See Figure 8.31.) The Error Out terminal of AI Waveform Scan is connected to the General Error Handler. With the value 2 for the type of dialog, a dialog box will appear if an error has occurred, and the user has the option to continue or stop the VI.

Storing the Voltage Samples

Often you will want to save the measured voltages in a disk file for later analysis, use by another program, or transfer to another computer. For example, you may want to acquire the data in the laboratory, then transfer it to another computer for further data reduction and plotting. You can transfer the data by writing it onto a floppy disk and carrying that disk to the other computer or by transferring the data file over a network. In either case, the first step will be to store the data in a disk file on your laboratory computer.

As LabView is working, it stores your data internally in a binary format. For example, your voltage measurements are normally stored as single precision floating point numbers. Floating point is the equivalent of scientific notation for a computer. Recall that in scientific notation the number 24,530,000 is written as 2.453×10^7. Single precision floating point numbers are stored as a set of 32 binary bits. The data in a disk file are not normally stored in this binary format but are instead stored as a sequence of characters. For example, 2.453×10^7 would be stored as the seven-character sequence 2.453e7 (the decimal point counts as a character). The characters themselves must be represented by a binary code, since data storage and data transmission devices can use only 1's and 0's. A common code called ASCII (American Standard Code for Information Interchange) is used for character representation in most computer-compatible devices. This code uses eight bits (one byte) to represent each character. For example, the character A is represented as 01000001. Therefore, our sample number requires 56 bits of storage (8 bits/character X 7 characters). It would be more efficient to store the binary numbers directly, but the formats for binary data storage are not standardized and you would likely run into problems in transferring data from one computer to another or even between different programs in your computer.

We next have to think about how the number is represented as a set of characters. We could represent the number 786 as 7.86e2, 786.000, or 786. Which one we choose is determined by a format specification. The format specification used by LabView is similar to that used in the C programming language. The range of possible formats is discussed under Format and Append in the LabView Function Reference Manual. You should normally use the d format for integer data or the e format for nonintegers. If you are not sure what to use, then choose the e format, which will convert the data to scientific notation. A typical e format statement is: %8.3e, which means that scientific notation should be used with three significant figures after the decimal point. The 8 here indicates the minimum number of characters or spaces to be used. The number 138.73 would be stored as 1.387e2 using this format. A very easy format statement is to write %e as shown in the following examples. This automatically uses the number of characters required to write the number. Our example would be stored as 1.3873e2 using the %e format.

We also have to think about another formatting question when an array of data is stored in a file. That is, how do we separate the elements of an array? There is no standard format but the format used by spreadsheet programs like Lotus or Excel is very commonly used. Many graphics programs and other applications expect the array data to be in spreadsheet format. Also, the data can easily be read into a word processing program. In spreadsheet format, the successive elements in one row of the array are separated by the tab character. Successive rows are separated by the EOL (End of Line) character. We recommend that you use this format for all of your data array storage.

Exercises 7 and 8: Storing and Retrieving Voltage Data

You should now make use of what you've just learned to develop a simple VI to acquire and store voltage data in a spreadsheet format. For Exercise 7, you should write a VI to acquire multichannel data and store it in a spreadsheet. The VI should be similar to your exercise 4, but rather than plot the data you should store them to a file. To confirm that this really worked, you should do exercise 8 in which you will write a VI to read the data from the file you previously wrote. This VI should then plot the data. Before proceeding with these exercises, you should go back to the LabView Tutorial manual and read Chapter 6 on Strings and File I/O. This will teach you about the VIs needed to format and store your data. Note as you read this that you do not have to store your data in the spreadsheet format. However, we recommend that you use this format. If you have a spreadsheet program on your computer, you should be able to load the measured voltage array into a spreadsheet. As usual, the exercises are worked out below, but try doing them yourself before using our solution.

The VI Vsave acquires a set of voltage samples from specified channels at a specified rate and stores them in a designated file. The front panel is similar to VseqP2, with extra terminals to control storing of the data (see Figure 8.32). Append to file and Transpose are Boolean controls. Set Append to file to "append" if you want to append the data to an existing file, or set it to "new file" to create a new file or to replace an existing file. The Transpose input specifies whether you want to transpose the array of data. The File path specifies where the file will be written, starting with the hard drive name and a colon, followed by folder names separated by a colon and ending with the file name. If you leave

70 Analog-to-Digital Conversion

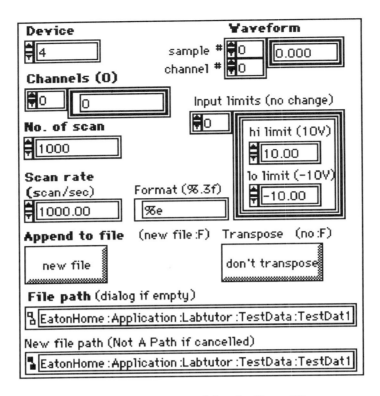

Fig. 8.32 Front panel for the Vsave VI

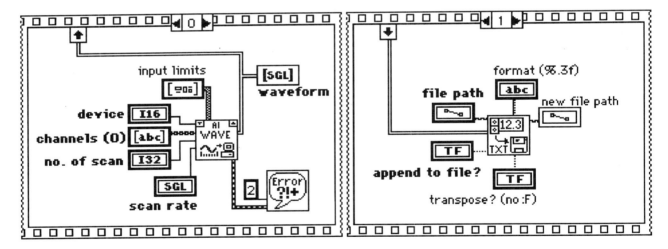

Fig. 8.33 Diagram for Vsave

this control empty, the VI will produce a dialog box to ask you for the file specification. This is the easiest way to do it. The New file path is an indicator of where the file was actually written. Format specifies how to convert the data to characters. This example specifies the simple %e format as discussed above. This will store the data written in scientific notation with as much precision as needed. For example, a voltage measurement of 0.537 volts would be stored as 5.37e-1.

The diagram for Vsave is shown in Figure 8.33. In frame 0 of the sequence structure, the voltage samples are collected. AI Waveform Scan VI is attached to its terminals, and the waveform is sent to the next frame. In frame 1, data are written to a file using the Write to Spreadsheet File VI. The Write to Spreadsheet File VI is located under the Function menu in the Utility submenu in the File palette.

The Plot File VI (see Figure 8.34) is our solution for reading a file and plotting the data on the front panel. The data are shown in the Waveform array. The File path and format are specified as in Vsave. Format should be the same format that was used to store the data. Number of rows is the maximum number of rows the VI reads from the file.

The diagram for Plot File is also shown. Frame 0 shows the Read from Spreadsheet File VI connected to its terminals. The voltage data are connected to the front panel indicator Waveform and also sent to frame 1. In frame 1 the data array is bundled together with the other inputs required for the graph and the information sent to the graph icon.

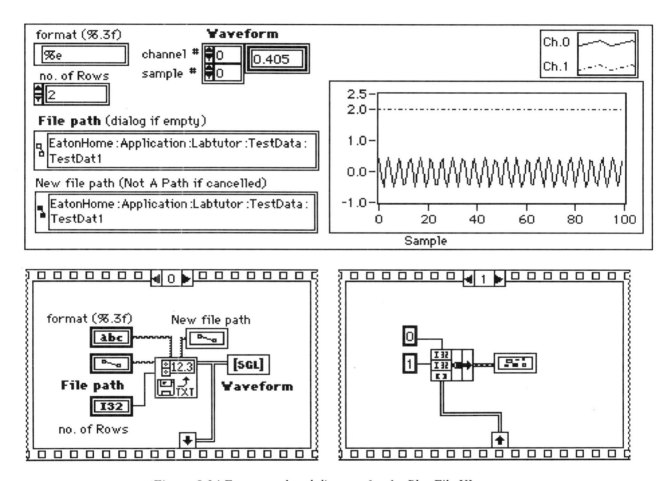

Figure 8.34 Front panel and diagram for the Plot File VI

Double-Buffered Data Acquisition

Double-buffered data acquisition is a technique for acquiring a very long sample record from the ADC and storing it in a disk file. It is used when the sample record is so long that it would overflow the available memory. A buffer in memory is repeatedly refilled by the data acquisition process while blocks of data are read from the buffer and written to the disk file. As long as the data acquisition rate is not too high, the data can be written to the disk file as fast as they are acquired by the ADC. If a fast data acquisition rate is used, a backlog of samples will build up in the buffer, and eventually the data acquisition process will overwrite data that have not yet been written to the file.

LabView supports double-buffered data transfers with several intermediate and advanced VIs. However, the use of these VIs is somewhat confusing. To make their use clearer, we have developed two VIs: ADDB, which acquires a data record of any length and stores it to a disk file, and DBGraph, which reads in a file acquired by ADDB and makes a graph on the screen. It would be easy to modify DBGraph to process the data record statistically or to modify the file in other ways.

ADDB is a modified version of a VI called Cont Acq to File (scaled), which is an example VI provided by LabView. Its function is to acquire samples continuously from the ADC and to write the samples in blocks to a disk file. The acquisition continues until you click Stop. The ADDB VI has been simplified and some of the controls removed to make it easier to understand. Although you can use ADDB VI to acquire data from multiple channels, the value for Number of Scans Written to File will be incorrect. Once you understand how this simpler version works you might want to use the more complete Cont Acq to File VI. Its file path is LabView 3:Example:DAQ:Analogin:STRMDISK:Cont Acq to File.

The panel for ADDB includes some new controls and indicators you haven't seen before (see Figure 8.35). The device, channels, and scan rate controls are the same as in previous VIs. The Buffer Size control is the number of scans the buffer will hold and is in units of scans. If you specify two channels and a buffer size of 1,000, the buffer will hold 2,000 data samples. The Number of Scans to Write at a Time specifies the size of the blocks that are read from the buffer and stored to the disk file. Scan Backlog is an indicator showing the number of scans in the buffer that have not yet been stored. If Scan Backlog increases continuously, it means that the VI is not reading data from the buffer fast enough and eventually the buffer will overflow. You can increase your buffer size, decrease scan rate, or increase the number of scans written at a time. The Number of Scans Written is an indicator showing the total number of scans written in your file. This VI continues to acquire and store data until the Stop button is pressed or an error occurs. You do not specify the file on the front panel. A dialog box will appear requesting a file name.

The diagram for ADDB is shown in Figure 8.36. Starting at the upper left, the Open/Create/ Replace File VI prompts the user to input the filename with the message

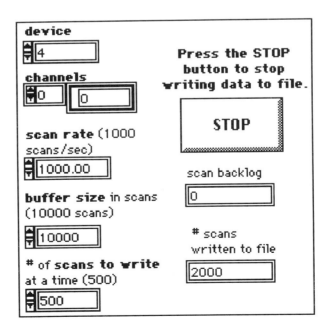

Fig. 8.35 Front panel for ADDB

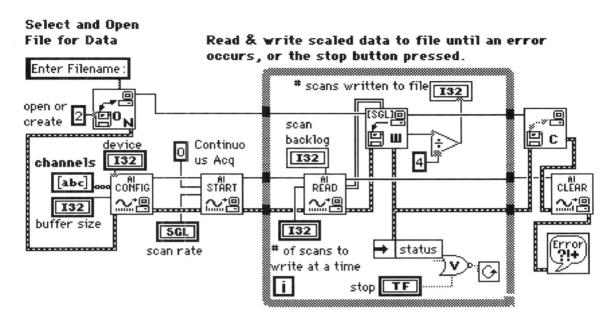

Fig. 8.36 Diagram for ADDB

"Enter Filename." The 2 connected to the Function input tells the VI to create a new file or replace an existing file. If the file name already exists, the VI will prompt you with a message to verify that you want the file replaced. Refnum is the reference number of the file and is connected from Open/Create/Replace VI to the Write File+ [SGL] VI and to the Close File+ VI. The connection from Open/Create/Replace to AI Config is Error Out. The Error Out from each VI is connected to the Error In terminal of the following VI throughout the diagram.

AI Config sets up the ADC operation so that it includes inputs for device number, channels, and buffer size. The input limits would also be specified here, but we have used the defaults to simplify the diagram. Task ID is connected from AI Config to AI Start to AI Read and finally to AI Clear. The AI Start VI starts the buffered analog input acquisition. Specifying 0 for the Number of Scans to Acquire input tells the VI to acquire samples continuously into the buffer until you clear the analog input task, using the AI Clear VI.

The data are read from the buffer and written to the disk file in the While Loop, which runs until the Stop button on the front panel is pressed or there is an error out of the Write File+ [SGL] VI. The Unbundle by Name function is used to isolate the status element of the Error Out cluster. The Not Or function computes the logical NOR of the inputs, stopping the While Loop when either of the inputs are true.

Within the loop, the AI Read VI reads a block of data from the buffer and determines the backlog of samples remaining in the buffer. The latter output is connected to the Scan Backlog indicator. The voltage data are sent from AI Read to the Write File+ [SGL] VI. Each voltage sample is represented as a single precision floating point number requiring four bytes of storage. Write File+ [SGL] VI writes the numbers to a byte stream file on the disk. The Write File VI also gives the number of bytes stored in the file. We use this output to compute the number of samples already stored in the file. The output must be divided by four since each sample occupies four bytes of storage. The Close File+ VI closes the file and AI Clear clears the acquisition task.

DBGraph (see Figure 8.37 reads the data from the disk file in blocks and plots the data as a strip chart. The only controls are Number of Scans to Read at a Time which sets the block size, and the STOP button. The VI prompts the user for a file name. The data

74 Analog-to-Digital Conversion

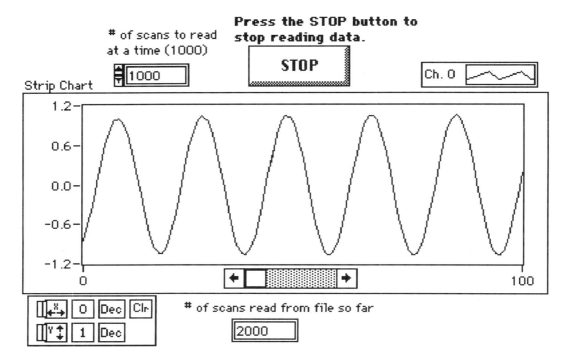

Fig. 8.37 Front panel for DBGraph

are displayed on a Waveform Chart with the strip chart and show palette options selected. Each time you run this VI, you need to clear the chart or the VI will plot the new data at the end of the old data. You can change the range of the x and y axes to show the data clearly. The indicator at the bottom shows the number of scans that have already been read.

DBGraph is a modified version of a VI called Display Acq'd File (scaled), which is an example VI provided by LabView. Its file path is LabView 3:Example:DAQ:Analogin: STRMDISK:Display Acq'd File (scaled). The DBGraph VI is simplified and has fewer controls compared to Display Acq'd File (scaled). Although you can use DBGraph VI to plot data from multiple channels, the value for Number of Scans Read from File will be incorrect. If you do use DBGraph for multiple channels you will need to expand the legend box.

The first and last parts of the diagram (Figure 8.38) open and close the file. The General Error Handler VI is connected to Close File+ VI. Within the While Loop, the Read File+ [SGL] VI reads blocks of data from the disk file. The output array is sent to the Strip Chart indicator. The number 1 connected to the Read File+ [SGL] VI specifies the number of columns for the output array. This is the number of channels in the file of data. If you use DBGraph to plot multiple channel data, you will have to change this.

The number of scans already read from the file is calculated by dividing the Mark After Read (bytes) output by four. The number four is used since it takes four bytes to represent one data sample. This calculation is only for a file with one channel. For multiple channels, divide the Mark After Read output by the product of the number of channels and four.

Three different conditions can stop the While Loop: you can push the STOP button on the front panel, the Read File VI can reach the end of the file, or an error can occur. The Unbundle by Name function is used to extract the status element from the Error Out cluster. The Or function then produces a true if the end of the file has been reached or an error has occurred. The Nor function produces a false signal, which stops the loop when any of the three stopping conditions is true.

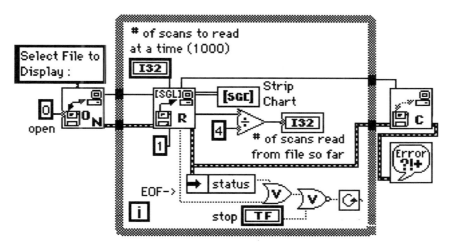

Fig. 8.38 Block diagram for DBGraph

Exercise 9: Testing Double-Buffered Acquisition

We are now ready to try the double-buffered acquisition VIs. First you should acquire some data with the ADDB VI. Set up a function generator to give a sine wave at about 20 Hz, and set ADDB to sample the signal at 200 samples/sec. Now start ADDB, and specify a simple file name that will be easy to remember. Let ADDB run for about 20 seconds, but while it is running twist the amplitude knob on your function generator back and forth. This way you will be able to see some variation in your plot. Be sure that you don't let ADDB run for too long at this rate, or you will create a huge disk file.

After you have stopped ADDB, start DBGraph and open your data file. Change the range on the plot to give a good picture of the data, and scroll back and forth to see the changes in amplitude.

When you are done with this exercise, delete the data file so that the disk doesn't get filled with useless files.

The continuous sampling VIs can be used for more than just double-buffered data acquisition. In some cases you might want to monitor continuously an input signal, but record it only when something interesting happens. An example would be a seismograph, which records ground motion during an earthquake by continuously storing the signal from an accelerometer, a device that produces a voltage proportional to the acceleration it experiences. Usually the ground does not move except for small vibrations caused by moving machinery or vehicles. The ground vibrations from an earthquake arrive at unpredictable times. If we wish to sample and store the signal caused by an earthquake, we must keep the ADC continuously running. This will produce an incredibly long disk file, most of which will be useless.

We will develop a new VI, which will keep the ADC continuously running while simultaneously monitoring the data. The data buffer will be repeatedly overwritten until the input level exceeds a preset trigger level. At that point the VI will jump to a second loop which will write data to a disk file for a specified amount of time.

The ADDBTrig VI is similar to ADDB in that data are acquired continuously, but the VI begins writing data to the file only when the acquired voltage samples exceed a specified trigger value. If you stop the VI before the trigger occurs, it creates an empty file.

The front panel is very similar to the panel for ADDB, with the addition of the Comparison Voltage control. (See Figure 8.39.) There are also two Scan Backlog indicators, since there are two AI READ VIs in the block diagram. This VI is written for a one-channel data acquisition. The sampling rate is set in the diagram, and default values are used for the input limits to keep the diagram simple. It would be easy to add front panel controls for these.

76 Analog-to-Digital Conversion

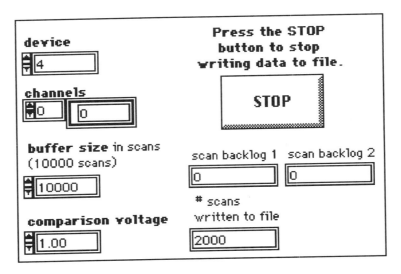

Fig. 8.39 Front panel for ADDB Trigger

The first part of the diagram shown in Figure 8.40 is similar to the diagram for ADDB. On the left, the file is opened and the ADC configured and started. The Scan Rate control is set at 1,000 scans per second but can be easily changed.

In the While Loop, the AI Read VI output is compared to the Comparison Voltage value. The AI Read VI reads only ten voltage samples during each cycle, so the output array dimension is ten rows by one column. Initialize Array, which is above the While Loop, creates an array with the dimensions of ten rows and one column to match the output array from AI Read VI. Each element of the array is set equal to the comparison voltage. The Greater Or Equal function compares the two arrays. The Compare Elements mode is selected, so the function compares individual elements from the arrays. Its output is a boolean array with a dimension of ten rows by one column. The output of Or Array Elements function will be true if any of the elements from the output of the Greater Or Equal function are true. Thus, the While Loop stops if any of the elements in the array output from AI Read VI are greater than or equal to the comparison voltage. The loop will

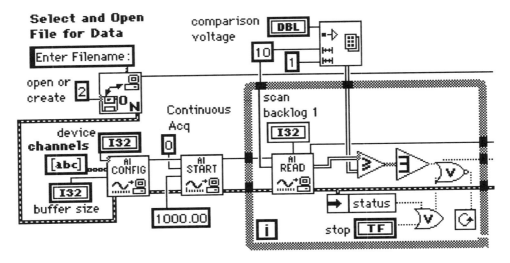

Fig. 8.40 The first part of block diagram for ADDBTrigger

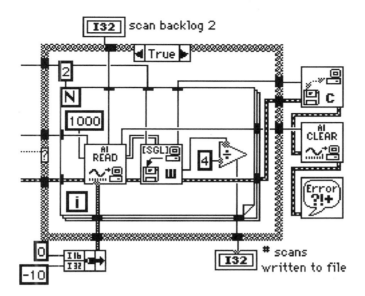

Fig. 8.41 The second part of block diagram for ADDBTrigger

also stop if you press the STOP button on the front panel or if an error occurs.

The Scan Backlog 1 indicator is for the AI Read VI in the While Loop. If this value increases while you run the VI, then adjust the Scan Rate, Buffer size, or the Number of Scans to Read as explained in the ADDB VI description.

If a trigger occurs, then the VI continues to the True subdiagram of the case structure shown in Figure 8.41. The True subdiagram has a For Loop, which contains AI Read and Write File+ [SGL] VI's. The For Loop executes 20 times and 1,000 scans are read and written to a file during each cycle. Therefore, 20 seconds of data are stored after the trigger event. These variables can be changed but were excluded from the front panel for simplicity.

A cluster is built below the loop and connected to the Read/Search Position input of AI Read. The cluster of Position and Read Offset values defines where in the acquisition buffer you want to start reading. The 0 for Position and -10 for Read Offset specify that the AI Read VI in the For Loop will start at a position 10 scans back from where the AI Read VI in the While Loop ended. The Read Offset is set to -10, so data where the voltage value initially exceeded the comparison voltage will be written to the file. The Read Offset may be changed if you would like to store more of the samples that occurred before the trigger. The Number of Scans Written to File is calculated by dividing Mark After Write (bytes) output from Write File+ [SGL] VI by four.

The False subdiagram is not shown. It assigns a zero value to the Number of Scans Written to File. The input tunnels for Task ID, Reference Number, and Error In are wired to their corresponding output tunnels.

8.7 Hardware Exercises

Testing ADDBTrigger

The first hardware exercise will be to test the VI we just developed. You are unlikely to be able to buy an accelerometer with sufficient resolution to sense earth movements so we will instead test the system using a microphone. Imagine that you are trying to record the sound of a bird singing in its nest. You hide a small microphone in the nest and connect it to the ADC on your computer. Most of the time the bird is away from its nest, and even when it is there it sings only sporadically. You could use ADDBTrigger to record the microphone signal only after the signal exceeds a threshold thereby avoiding storing many millions of useless samples.

You can test this with an actual bird if you know what you are doing but we recommend testing it in the laboratory. A handheld microphone like those used by singers can be purchased for around $25 at many electronics stores. We recommend a Realistic Super Cardioid microphone or equivalent. Most microphones have an RCA phono jack connection so you will need to make an adapter to wire it to the BNC inputs on your ADC. The microphone will produce a signal of around 20 millivolts peak to peak with normal sound levels so you will have to adjust the input limits on your ADC. You do that by connecting an input limits cluster to the appropriate input on the AI Config VI. Look at the signal from your microphone on the oscilloscope before setting the input limits. You will also have to change the comparison voltage control to a smaller value so that the microphone signal can be sensed by the triggering loop.

Once you have modified ADDBTrigger appropriately, start it up. Make sure you are in a quiet room so that the data recording won't be triggered immediately. After waiting for a few seconds start singing or whistling. You should see the value of Scans Written to File start to change when the actual triggering occurs. If you cannot make the VI trigger, you should change the comparison voltage to a smaller value. After the VI completes storing the file, you can examine the data recorded using DBGraph.

A Simple Control Loop

A lab computer is sometimes used as a feedback element in a control loop. A typical control loop is shown schematically in Figure 8.42. One or more voltage samples is read in through the ADC, processed appropriately; then a control voltage is sent back out through the DAC. For example, a closed-loop controller could be made for the DC motor used in the last chapter. A tachometer would be required to sense the motor speed. The computer would read the tachometer output, compute the difference between the desired and actual speeds, and then change the motor drive voltage appropriately.

A problem in using a lab computer in a control loop is the response time required. A standard computer is generally adequate as a controller for systems requiring response times of the order of 0.1 second or greater, with the lower limit of the response time being determined by the specific computer used. Much faster control loops can be implemented on personal computers with the addition of some special purpose interface boards.

For this exercise, you will test the response time of your computer by developing the simplest possible control loop: one that reads a voltage in the ADC, then sends the same voltage back out via the DAC. Note that, normally, the sensor data would be processed before the control signal is sent. The first step is to set up the control loop using LabView. You should use AI Read One Scan and AO Write One Update embedded in a For-Loop structure. Make the number of cycles through the loop a front panel control. You will also need controls for the device number, input channel, output channel, input limits, and output

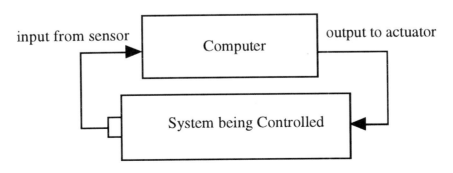

Fig. 8.42 Schematic of a simple control loop

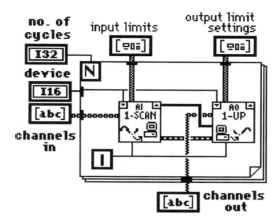

Fig. 8.43 Sample diagram for the control loop exercise

limits. This exercise is worked out below, but you should try to program this one yourself. Once you have developed your own VI, read on to learn how to test it.

The diagram in Figure 8.43 shows the voltage output from AI Read One Scan connected directly to the input control of AO Write One Update. The iteration counter is connected to both AI Read One Scan and AO Write One Update. This is done so that the ADC and DAC are configured only once instead of on each loop iteration. Arrays and clusters connected to terminals in a For Loop or a While Loop have the Enable and Disable Indexing option. You need to select Disable Indexing for Channels In and Out and Input and Output Limits. This is done by first wiring the desired input or output to its terminal. If a dotted line appears, indicating that the wiring was not successful, command click at the input tunnel on the loop border and select Disable Indexing.

When Enable Indexing is selected for output tunnels, the output values are accumulated using an additional dimension. For example, the Voltage Out from AI Read One Scan is a one-dimensional array: a single sample for each channel. To wire the voltage output terminal of AI Read One Scan to an indicator located outside the For Loop, you must create a two dimensional array. When the Voltage Output wire exits the loop, there is a row of data for each cycle of the loop. With each iteration, the new data are displayed by adding a row at the end of the existing array. After the first iteration, the data are displayed in row 0, columns 1-n, where n is the number of channels. After the second iteration, the array has two rows, and so on. The indexing of the array happens automatically whenever Enable Indexing is selected for the output tunnel. At the input tunnel, the opposite process occurs. A one-dimensional array will enter the loop one element at a time, and a two-dimensional array will enter one row at a time.

You will test the loop response time using a function generator and an oscilloscope. (See Figure 8.44.) Set up the function generator to supply a triangle wave output with about 2 volts peak to peak amplitude at a frequency of 40 Hertz. Make sure that the signal does not contain significant DC offset (that is, the waveform should be centered around 0 volts). Now set up the oscilloscope for dual-trace operation. Set the scope in Norm trigger mode with a sweep rate of 5 msec/div. Both inputs must be DC coupled. If you do not see the wave, try adjusting the trigger level. You should observe one triangle wave and a stair-step wave, which is offset in time. The stair steps are likely to be moving. You can make them hold still by adjusting the input frequency slightly. The length of the stairs tells you how long it takes the loop to respond. Decrease the frequency, and the output wave will begin to look more like a triangle wave. At higher frequency, the output will be a very poor representation of the input.

80 Analog-to-Digital Conversion

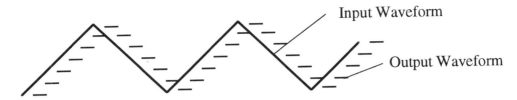

Fig. 8.44 Oscilloscope trace during test of loop response

A Faster Control Loop

The control loop seems to run very slowly. Part of the slowdown is that AI Read One Scan converts the raw binary reading to a true voltage. AO Write One Update then reconverts the voltage to a binary value for conversion. Set up a new control loop in which you eliminate the conversion of the data. AI Read One Scan has Output Units control, which allows you to return binary data only, but AO Write One Update does not accept binary data. You can fix this by modifying AO Write One Update slightly. Open up the block diagram of AO Write One Update. The AO Single Update VI, located inside the While Loop, has a terminal for binary data. Save a copy of AO Write One Update, using a different name and using the Save A Copy As command under the File menu. We renamed it AO Write One Update Control Loop2, as shown in Figure 8.45. Add a binary array input terminal to the icon of your renamed AO Write One Update and the front panel of your new Control Loop VI. Be sure that you do not modify the original AO Write One Update VI. Replace the AO Write One Update VI with the modified version in your control loop VI as shown in Figure 8.46. Note that you must wire a constant value 2 to the Output Units control on AI Read One Scan. This tells the VI only to return binary data.

When you are done modifying your VI, retest the control loop. Is it significantly faster? Try varying the input frequency to see how high you can go and still have a reasonable-looking output waveform.

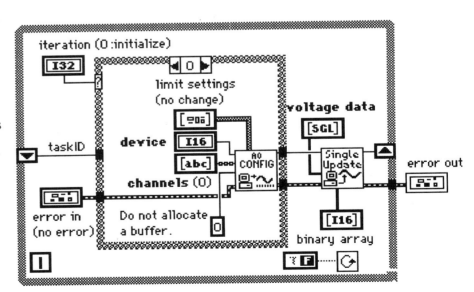

Fig. 8.45 Diagram from AO Write One Update Control Loop2

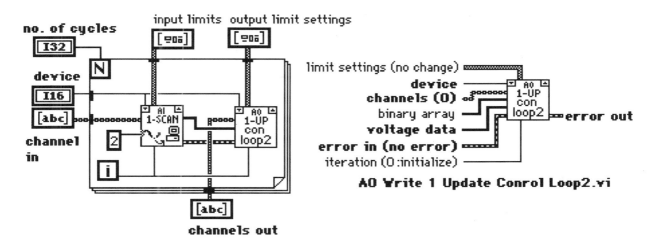

Fig. 8.46 Diagram for the faster control loop and the icon for the modified AO Write One Update

Even the faster control loop we just developed is way too slow for many applications. The reason that it is so slow is that our ADC and DAC cannot run simultaneously. We are repeatedly executing a subprogram to configure the ADC, then acquire a voltage sample. We then reconfigure the board to use the DAC and send out the voltage. All of these steps are repeated at each iteration of the loop. To use the computer for a control loop operating at higher frequency, we need to read in voltages continuously from the ADC under DMA control, storing the results in a buffer. We then perform calculations on the samples in the buffer to determine the appropriate output voltages. These are then loaded into a second buffer. The DAC, also operating under DMA control, sends out voltages from the second buffer.

Control loops such as the one described above can be implemented in personal computer systems. Somewhat more sophisticated interface boards are required, and the programming is more complex. But, now that you have mastered the concepts presented here, you should be ready to tackle this more difficult problem.

Chapter 9

Parallel Digital Interfacing

9.1 Introduction to Digital I/O

Digital Input/Output (I/O) means that a single conductor can carry only a single bit of information at a time; that is, the applied voltage is either low or high and is interpreted as either a 1 or a 0. Contrast this to analog communication, where a single conductor may be used to transmit a voltage such as 4.327 volts, which contains about 13 bits of information. A digital communication link to transmit that same number would require a total of 14 conductors (13 bits + 1 ground) or else the ability to send the number one bit at a time and reassemble it at the receiving end.

An important property of digital communication links is their noise immunity. An analog signal cannot be communicated accurately in the presence of significant electronic noise. On the other hand the noise has to be very large to interfere with a digital communication link where the receiver has only to differentiate between high and low voltage levels.

Digital interfacing techniques are divided into several different types. In serial interfacing, data bytes or words are disassembled into individual bits, which are sent one at a time over a single conductor (or optical pathway). The receiving device must be capable of reassembling the bits in the correct order in order to make sense of the data transmitted. In parallel interfacing, several bits (usually 8 or 16) are communicated simultaneously over multiple conductors. Within the category of parallel interfacing, there are several subcategories. The GPIB is a standard parallel digital interface used for communicating with intelligent instrument systems. (We will learn about GPIB in Chapter 10.) Special purpose parallel interfaces are designed for high-speed communication between a computer and a peripheral device. Normally, the user has little opportunity to modify such an interface, so we will not study it. General-purpose parallel interfacing used in connecting a wide range of digital transducers to the computer is the topic of this chapter.

Most general-purpose data acquisition boards contain a digital I/O section (also called a parallel port), which serves as the interface between external digital devices and the computer's bus. The parallel port usually contains 8, 16, or 24 bits, which are configurable in blocks of 8 or 4 bits. The diagram in Figure 9.1 shows an eight-bit parallel port configured as an input port. The digital inputs are connected to latches that capture and hold the input digital value when triggered by the board controller. The set of input bits is then passed to the computer's bus as an eight-bit number. When the port is configured for output, the latches change their output state when the port receives a new number from the bus. The digital outputs are held until a new number is received from the processor via the bus.

The parallel digital interface is used for a very wide variety of applications which, can be classified into three major groups: single-bit devices, inherently digital transducers, and specialized instruments. A table listing some example applications is shown in Figure 9.2.

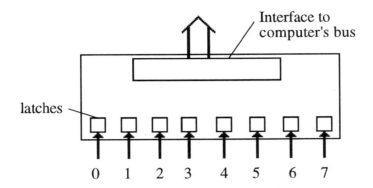

Fig. 9.1 Schematic of a general purpose parallel port configured for input

Single-bit devices	Switch Proximity detector Light detector	Light emitting diode (LED) Relay Transistor switch
Inherently digital transducers	Counter Digital encoder (used to indicate linear or angular position)	Stepper motor
Specialized instruments	Many specialized scientific instruments use custom parallel interfaces. The user must use the general-purpose parallel port to communicate with such instruments.	

Fig. 9.2 Example applications of a general purpose parallel interface

9.2 Digital Interface Hardware

Parallel interface ports for computer interfacing applications use TTL (Transistor/Transistor Logic) levels for the input latches. You must ensure that all digital devices connected to the port use compatible signal levels. Fortunately, most digital transducers are also set up to use TTL signal levels. The nominal TTL signal levels are 0 volts for the low state and 5 volts for the high state. The standard allows considerable latitude. To output a high state, a device must set the voltage on the line to a voltage above 2.4 volts. A receiver will interpret any voltage above 2.0 volts as a high input. The difference of 0.4 volts is called the high-state noise margin, as illustrated in Figure 9.3. To output a low state, a device must hold the voltage on the line to a level below 0.4 volts. An input device interprets any voltage below 0.8 volts as a low input. The difference in these levels is called the low-state noise margin. The voltage levels mentioned above are appropriate for standard TTL logic. Many devices use Low-Power Schottky (LS) series TTL logic. The nominal logic levels for this series are the same, 0 and 5 volts, but the noise margins are slightly different, as shown on Figure 9.3.

The high state does not necessarily correspond to a logical 1. In negative-true logic the

84 Parallel Digital Interfacing

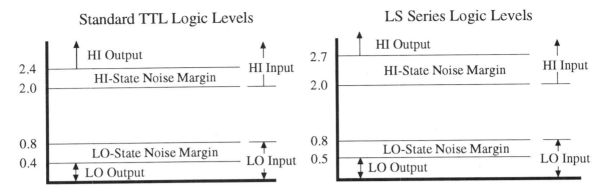

Fig. 9.3 Logic levels for two different types of logic circuits

Logic Sense	Logical Value	Output State	Nominal Voltage
Positive True	1	High	5 Volts
	0	Low	0 Volts
Negative True	1	Low	0 Volts
	0	High	5 Volts

Fig. 9.4 Negative and positive true logic definitions

opposite is true; that is, the high state corresponds to logic 0 and the low state corresponds to logic 1. It is important to note that the signal levels are the same for both positive- and negative-true logic. Negative voltages are never used in TTL interfacing. (See Figure 9.4.)

The parallel input ports used for computer interfacing typically use TTL logic with so-called open-collector inputs. The details of an open-collector input are not important but you must understand the current flows involved. The sketch in Figure 9.5 shows a simple switch input to 1 bit of a parallel port. When the switch is open, the input will go to the HI state. To pull the input to the LO state, the switch must be closed, connecting the input to ground. Up to 1.6 mA of current will flow out of the input pin when the switch is closed. Therefore, an output device connected to the input must be capable of sinking at least 1.6 mA of current.

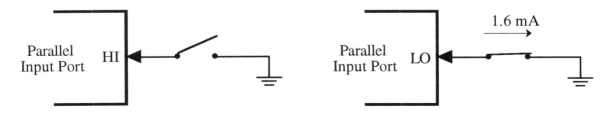

Fig. 9.5 Illustration of current flow during input to a parallel port

Digital Interface Hardware **85**

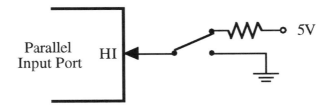

Fig. 9.6 A better input connection for a parallel port

Figure 9.5 indicates that you can just leave the input disconnected to input a high value. This usually works, but it is risky. The open input acts as an antenna and may pick up enough noise to cause the input to go to the low state. A better setup for connecting a switch is shown in Figure 9.6. The 5-volt source guarantees that the input will stay in the high state, and the resistor regulates current flow into the input. Even this type of input can have problems because the contacts of a normal switch will bounce (disconnect and reconnect) several times over a short duration as you change the switch position. This is called switch contact bounce, and it can cause problems in some applications. To avoid this problem, you can build a simple switch "debounce" circuit, described in many electronics textbooks.* Of course, any standard TTL output device will also work with the parallel input port.

When the I/O port is configured in the output mode, the user must be concerned with its output characteristics. A TTL output is designed to be connected to 10 or more TTL inputs. Therefore, when the output is in the low state, it must be capable of sinking up to 16 milliamps. On the other hand, the current sourcing capability in the high state is typically only about 400 microamps.

Connecting Some Common Devices to a Parallel Port

A light emitting diode (LED) is a convenient device to use as an indicator light and sometimes as a light source for instrumentation purposes. It can be connected to the parallel output port using a 7406 inverting buffer as shown in Figure 9.7. The 7406 is an

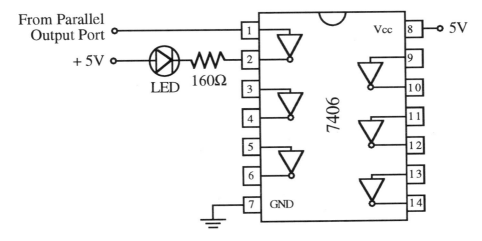

Fig. 9.7 Connection of a light emitting diode to a parallel port

---

* A good description may be found in *The Art of Electronics* by Horowitz and Hill (Cambridge University Press, 1989).

inexpensive integrated circuit containing six inverters. An inverter is used to buffer the output and to make it so an output high state corresponds to the LED turned on.

A parallel output port cannot supply sufficient current to drive any device with significant power dissipation. You use a relay controlled by a parallel output port to turn on/off lights, motors, and other devices. The relay itself draws more current than the parallel port can supply. Therefore, the port is used to control a power transistor, which in turn controls the relay. (See Figure 9.8.)

Some special relays can be connected directly to a parallel port without using a power transistor. We use a photo-isolated solid-state relay (Magnecraft Model W6202DSX-1) purchased from Newark Electronics of Chicago. This relay can switch regular AC line voltage (110V) at currents up to 2.5 A with its control input connected directly to a digital I/O port. The diagram in Figure 9.9 shows a schematic of this relay connected to turn on a standard 60 W light bulb.

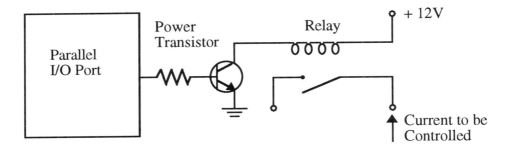

Fig. 9.8 Parallel port controlling a relay using a power transistor as a buffer

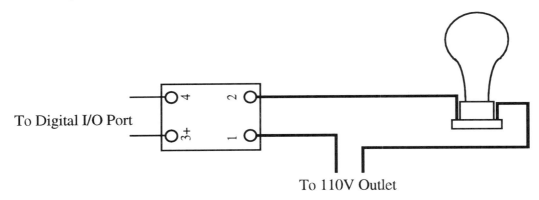

Fig. 9.9 A schematic of a relay connected to control power to a light bulb

9.3 Parallel Interfacing Protocols

A parallel port is frequently used to connect a digital device to the computer, allowing communication of data from the device to the computer and vice versa. Various communication protocols are used, depending on the application. A protocol is a set of conventions governing how data will be communicated. The protocol may determine how the data are formatted and how the data transmission is controlled.

The simplest protocols, processor-controlled (untimed) and processor-controlled (timed), are not really protocols at all. The external device is essentially passive and need know nothing about the protocol. More complicated protocols, including handshaking and

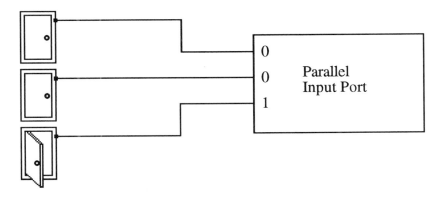

Fig. 9.10 A typical setup requiring the processor controlled untimed protocol

interrupt-driven communication, involve cooperation between the computer and the external device to control the data communication.

Processor-Controlled (Untimed): In the processor-controlled (untimed) protocol, the data are assumed always to be valid, so the port can be read at any time. The programmer need not pay any attention to timing. An example of this protocol is status-checking used in a computerized burglar alarm. Imagine a setup in which every door and window in a building has a switch indicating whether the door is open or closed. Each switch is connected to one bit of a parallel input port. The processor periodically reads the input digital number to determine if any door has been opened. (See Figure 9.10.)

Processor Controlled (Timed): This is a somewhat unusual protocol and a risky one to use. The communication is still controlled entirely by the processor; however, the data are not always valid. The processor must keep track of time to ensure that valid data are available when the receiving device reads the data. An example is an old-fashioned instrument that performs an operation (say, measuring a voltage) on receiving a strobe signal from the computer. The actual measurement takes some time to complete, so valid data will not be available on the communication lines until a set time interval after the strobe has passed. The programmer must be aware of the time delay and account for it when designing the communication software.

A better way to deal with this is to have the external device tell the processor when the data are ready. This is one of the ideas behind the more complicated protocols described below.

Handshaking: Handshaking is a protocol in which the computer and the external device jointly control the communication. Imagine a situation in which an external device is communicating a stream of data bytes (say, successive measurements) to the computer. We want to make sure that the computer reads every data byte and does not read any twice. Therefore, the external device must tell the computer when valid data are available. The computer in turn must tell the device when it has accepted the data and when it is ready for more. (See Figure 9.11.)

Handshaking is implemented using two additional one-bit lines, one from the computer to the device and one from the device to the computer. There are various ways in which

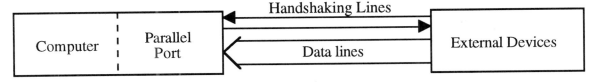

Fig. 9.11 Handshaking connection between a parallel port and an external device

88 Parallel Digital Interfacing

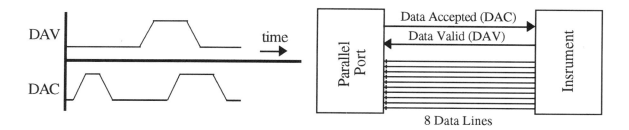

Fig. 9.12 Handshaking protocol using Data Valid and Data Accepted lines

handshaking can be done. Two common techniques are described here.

Consider an instrument transmitting data to the computer over eight data lines with Data Valid and Data Accepted handshaking lines. DAV high means that valid data are available on the data lines. A transition from low to high on DAC means that data were accepted, and a transition from high to low means that the parallel port is ready for more data. The parallel port must first set the DAC line high, then change it to low to initiate data acquisition by the instrument. The instrument acquires the data and then sets the DAV line high. The parallel port reads the data and sets DAC high, indicating that it has the data. The instrument then lowers the DAV line and begins to acquire a new data byte. The parallel port then passes the data byte to the computer processor over the system bus. When the parallel port is ready to accept more data, it lowers the DAC line which re-initiates the cycle. This protocol is illustrated in the timing diagram in Figure 9.12.

We now consider a different handshaking arrangement using lines called Strobe and Input Buffer Full (IBF). Usually the word strobe means that the line changes state in only a short pulse. For example, the instrument might normally hold the strobe line in the low state. When the instrument is ready to transmit data, it quickly pulses the Strobe line to the high state. The strobe may last for only a few microseconds, so the parallel port must be prepared to sense the strobe. In this arrangement, the parallel port uses a line called Input Buffer Full (IBF) to indicate when it is ready to receive data. In the example illustrated in Figure 9.13, the instrument first pulses the Strobe line, indicating that data are ready. The parallel port latches the data and then sets the IBF line. The data are then sent from the port to the processor, which clears the port to accept another data word. At this point the port clears the IBF line. The instrument then sends another strobe, beginning the cycle again. Note that in many cases the Strobe line is normally in the high state and pulses to the low side. This is generally be indicated by placing an overbar over the word Strobe, as shown in the example.

Interrupt-Driven Communication: Interrupt-driven communication is a protocol in which the external device controls the communication. The device issues an interrupt request to the port when it has data to be transferred. The port then interrupts the processor

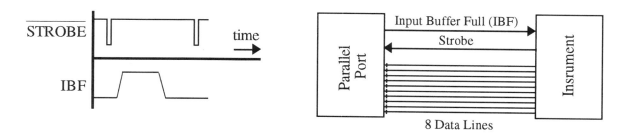

Fig. 9.13 Handshaking with strobe and IBF lines

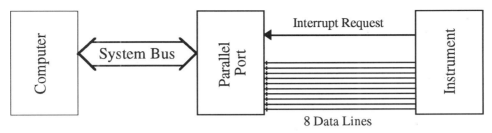

Fig. 9.14 Interrupt-driven communications

via the interrupt request line of the system bus. (See Figure 9.14.) The interrupt service subroutine then controls the port to read the data from the external device. Multiple data words may be transmitted following a single interrupt request. In that case, a handshaking system may be used to make a reliable communication link.

Pitfalls and Solutions: The most common and easily detected pitfall in parallel digital interfacing is disconnected signal lines. If the line connecting a signal source to a digital input is disconnected or broken, the digital port will always indicate a high-state input. This is easy to detect if you remember that an open input will settle at the high state.

Another common problem is inverted logic; that is, the external device is using negative true logic, while the interface board is using positive true logic. If you are successfully transferring data but the data don't make sense, you should check for this problem simply by treating 1s as 0s in your program and vice versa.

There are many pitfalls with handshaking connections. Some of the most common are inverted strobes, strobes that are too fast for the receiver to detect, and problems initiating data transfers. The first thing to do is to check the timing diagrams for compatibility. If a strobe is too fast, you may need to use a Schmitt trigger to capture it. A Schmitt trigger is easy to hook up if you know any electronics. If not, ask for help.

9.4 Parallel Interfacing with LabView

The NB-MIO-16 board we are using has fairly limited digital I/O capabilities. There are eight I/O lines, organized in two groups of four bits each. Each group may be configured for either input or output. The two four-bit groups are called ports, and are named DIOA and DIOB. The LabView digital interfacing VIs refer to these as Port 0 and Port 1 respectively.

The digital I/O part of the NB-MIO-16 is quite useful for controlling relays and other single bit devices but is not useful for communication applications where handshaking is required. It does not support automatic handshaking, and since there are only eight lines it would be difficult to set up your own. National Instruments sells other boards that are dedicated entirely to digital interfacing. These boards support automatic handshaking for both input and output. The boards are also supported by LabView but will not be described here.

The individual bits of the digital I/O port may be addressed as single bits or in groups of four. The four bits of each port are numbered 0 to 3. When addressing a group, the pattern of the four bits is represented as an integer number in the range 0 to 15. Two examples are shown below. When using the Easy I/O VIs you don't have to worry about this. The bit pattern is actually given as a sequence of 1s and 0s.

The bit value 1 corresponds to a TTL high state, and 0 corresponds to the TTL low-state; that is, positive-true logic is used. Many devices you may connect to the digital port use negative-true logic. This is easy to account for when you do your programming as long as you are aware of it. Just be sure that you associate a 1 value in LabView with a 0 value at the external device.

bit number	3	2	1	0	3	2	1	0
bit pattern	1	1	0	1	0	1	0	0
integer representation			13				4	

Fig. 9.15 Representation of the digital values by LabView

The Easy I/O VIs are all you really need to use the parallel port on the NB-MIO-16 board. We describe them briefly here, then show the front panels below.

Read From Digital Port: This VI is used to read in the bit pattern on the selected port. The output shows the bit values for each line of the port. You can set it up to read both ports simultaneously.

Read From Digital Line: The VI reads a single input line. For example you can read line 1 of port 0. The output is a Boolean true or false. The VI might often be used in triggering operations.

Write To Digital Port: The VI sends a four-bit pattern to a single port or an eight bit pattern to both ports. The pattern remains unchanged until the next output operation.

Write To Digital Line: This sets the selected line to the value indicated by the front panel switch. The other lines of the port are not affected. You use this to control single-bit devices.

The Read From Digital Port VI

The front panel for Read from Digital Port is shown in Figure 9.16. This VI reads the bit pattern for a specified port. For NB-MIO-16 board, the Port Number can be 0 or 1. If you specify 0 for the Port Number, Port Width can be 4 or 8. If you specify 1 for Port Number, then Port Width must be 4. Specifying 0 for the Iteration tells the VI to call a configuration VI, which sets up the specified port as an input port. The front panel shown here is displaying a bit pattern set by the eight switches on the computer tester. It indicates that bit 1 is 1 and bit 0 and bits 2 through 7 are 0. Notice that bits 2 through 7 are not displayed since they are all 0. If bits 7 and 1 are set to 1 and the rest of the bits set to 0, then the Pattern indicator would show 10000010.

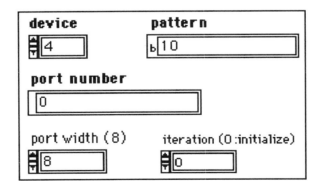

Fig. 9.16 Front panel for the Read from Digital Port VI

The Read from Digital Line VI

The front panel for Read from Digital Line is shown in Figure 9.17. The controls are similar to the controls for Read from Digital Port. The Line specifies the individual line or bit within the port that you want to read. Line State indicates whether the specified line is High (1) or Low (0). The front panel shown indicates that the line is at a high state.

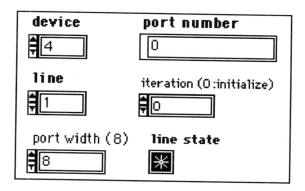

Fig. 9.17 Front panel for the Read from Digital Line VI

The Easy I/O VIs for Writing to the Parallel Port

The front panels for Write to Digital Port and Write to Digital Line are shown in Figure 9.18. Write to Digital Port is used to send a pattern to either of the four-bit ports or to both of them simultaneously as shown here. Write to Digital Line sends the value set by the line state switch to the selected line. If port number is set to 0, and port width to 8, then the eight output lines are numbered 0 to 7. If the port number is set to 1, then the port width must be 4 and the lines are numbered 0 to 3.

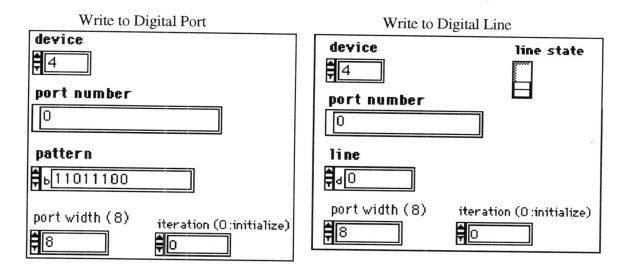

Fig. 9.18 Front panels for Write to Digital Port and Write to Digital Line

Exercise 1: Reading the Switch Register

In this exercise you will use the Computer Tester to check that you understand the function of the digital input VIs. The set of switches on the Computer Tester are used to supply input values to a set of eight digital lines, which can be read through the input port. Set up the Computer Tester in DIG IN Mode. Connect the cable from the DIGITAL connector on the computer interface box to the INPUT TEST connector on the back of the Computer Tester. Set a bit pattern using the switches labeled 0 to 7 on the front of the tester. Now you are ready to test the digital input VIs. Use Read from Digital Port to acquire the digital values. Change the switch values, and read the input port repeatedly until you are sure you understand which switch is connected to which bit.

Exercise 2: Controlling the LED Array

Now you will test the digital output capability of LabView and the NB-MIO-16 by using them to turn off and on a set of light-emitting diodes (LEDs). Each of the eight LEDs on the Computer Tester may be turned on by sending out a logical true (high-state output) and may be turned off by sending out a logical false. When properly connected, LED 0 on the Computer Tester is controlled by Port 0 Line 0, LED 1 by Port 0 Line 1, and so on.

Set up the Computer Tester in DIG OUT mode, and connect a cable from the DIGITAL connector on the computer interface box to the OUTPUT TEST connector on the back of the computer tester. Use Write to Digital Port to send out various bit patterns, and observe the results on the LED Array. Next, use Write to Digital Line to change single-bit values. Are the other bits affected when you change one of the bit values?

Exercise 3: Square-Wave Generator

In many laboratory applications you need to send out a square-wave train where you know precisely the number of cycles sent. An example is the control of stepper motors where the motor steps forward once for each square-wave cycle received. Develop a virtual instrument in which you send out a series of alternating 1s and 0s using a single line of the digital I/O port. The front panel should allow you to choose the number of cycles and which line and port to use. This exercise is worked out in Figure 9.19, but don't peek until you try it. Run the VI, and observe the signal on the oscilloscope. What is the square-wave frequency? Does the output spend as much time in the high state as in the low state?

Our solution is shown below. The diagram is just a For-Loop with an embedded two-frame sequence. Write to Digital Port is inside both frames of the sequence structure. The Iteration terminal is wired to the iteration terminal of the VI. The second frame of the sequence is exactly the same as the first, except that the binary constant is set to false.

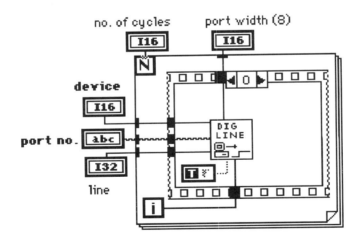

Fig. 9.19 Diagram for the sample solution to Exercise 3

A big problem with this VI is that you can't control the frequency, as you'll see when you test it. If you really want to make an accurate square-wave generator, it would probably be better to use the counter-timers available on the NB-MIO-16. We can slow down the square-wave generator by adding a wait function, the VI that looks like a metronome in the Dialog & Date/Time Menu. This allows you to program a wait where each wait cycle requires 1/60 of a second to complete. Try putting the metronome in various places, and see how the frequency and waveform change. For example, you could put one in each frame of the sequence as shown in Figure 9.20, or put one outside the sequence structure but still within the For Loop. This is one way you could create a programmable square wave, but you wouldn't have much frequency resolution.

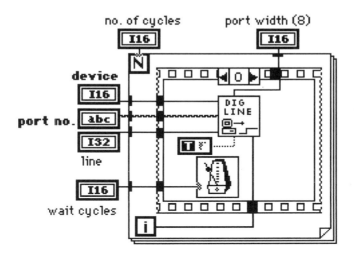

Fig. 9.20 Diagram for the alternative solution to Exercise 3

Chapter 10

The IEEE-488 General Purpose Interface Bus

10.1 Introduction to the GPIB

The IEEE-488 bus, which is commonly called the General Purpose Interface Bus (GPIB), is a special-purpose parallel bus that can be used to interface a wide range of instruments to small computer systems. The bus arrangement means that several instruments can be connected to the computer through a single plug-in card. The hardware connections and the communication protocols are all standardized, so it is relatively easy to connect a new instrument to the bus.

The General Purpose Interface Bus is used to connect the computer to many different devices, including standard laboratory instrumentation, special-purpose equipment, multi-instrument systems, and computer peripherals. Probably the most common use of the bus is to interface a collection of standard instruments to the computer in building a custom instrumentation system.

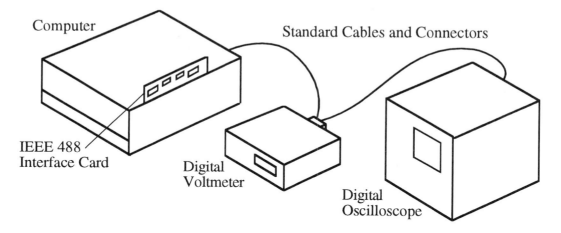

Fig. 10.1 A typical setup showing a computer connected via GPIB to two standard instruments

Laboratory Instruments	Computer Peripherals	Special-Purpose Systems
Digital Multimeter	Printer	Blood Chemistry Analyzer
Digital Oscilloscope	Plotter	Gas Chromatograph
Frequency Counter	Image Scanner	Engine Test Controller
Relay Scanner		Signal Processing System

Fig. 10.2 Some devices that can be connected to the GPIB

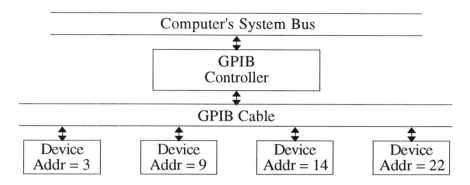

Fig. 10.3 Logical Organization of the GPIB

The logical organization of the bus is shown in Figure 10.3. Up to 14 devices are connected by a set of cables that form the shared communication pathway. Each device has a unique address, ranging from 0 to 30. The GPIB controller is plugged into the main computer system bus. It manages the activity on the GPIB bus and also communicates with the computer processor. Any device connected to the GPIB bus can be either a talker or a listener as designated by the controller under software control. Usually communication is between the controller and one of the devices, but the controller may designate one device as a talker and one or more other devices as listeners.

10.2 GPIB Hardware

The IEEE-488 Bus hardware consists of a bus controller card that plugs into your computer, standard cables to link the controller to the instruments, and an interface built into the instrument. There are very few options in purchasing the hardware. The controller cards from different manufacturers are all very similar and the cables are identical. The instrument interface is purchased as an option with the instrument. It is very easy to connect the cables, and it is impossible to connect them incorrectly.

The user really needs to know almost nothing about the hardware, other than the limits on the number of instruments that can be connected. Thus, most of this section is really just for information purposes and can be skipped if you want to get a quick start with the IEEE-488 Bus.

The controller is connected to the instruments via standard cables, which are terminated by "daisy-chain" connectors as shown in Figure 10.4. The daisy-chain connector allows the user to attach a new cable to the back side in order to extend the bus system. Up to 14

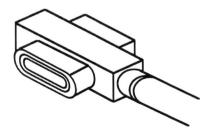

Fig. 10.4 The standard GPIB cable connector. The front side is plugged into your instrument, and another connector may be plugged into the back side.

devices may be connected to a single controller. To add an additional instrument to an existing bus system that has fewer than 14 devices attached, just plug a new cable into the back of any daisy-chain connector on the bus and extend the cable to the new instrument.

The bus may be arranged in a daisy-chain, a star, or a random topography as shown in Figure 10.5. There is no difference in performance, so select your topography for convenience of wiring. The bus does not need to be terminated.

The bus includes 16 one-bit signal lines listed in Figure 10.6. Eight of them are the actual data lines. All data are passed as eight-bit ASCII characters. Five of the lines are for control. Four of these are used by the controller to control the communication along the bus. The fifth is the service request line, which can be used from any device on the bus to request service. The last three lines are handshaking lines. Three handshaking lines are needed rather than the usual two because there may be more than one device receiving the data communicated. If you purchase good software you should not have to think about any of the signal lines, other than the data that are passed on the eight data lines.

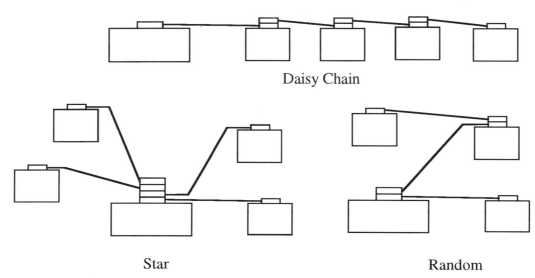

Fig. 10.5 Daisy chain, star, and random arrangements of the bus cabling

Control Lines	Handshaking Lines	Data Lines
ATN Attention	DAV Data Valid	8 lines
SRQ Service Request	NRFD Not Ready for Data	
EOI End or Identify	NDAC Not Data Accepted	
IFC Interface Clear		
Ren Remote Enable		

Fig. 10.6 The 16 signal lines of the GPIB

10.3 Transmitting Information on the GPIB

Information conveyed along the bus may be either commands or messages. Typical commands and messages are shown in Figure 10.7. Commands are instrument-independent (that is, they are the same for any instrument you purchase) and are always conveyed from the controller to an instrument. Messages are instrument-dependent and may be configuration data sent from the controller to the instrument or data sent from the instrument to the controller. The commands are communicated as eight-bit codes and the messages as sequences of characters. Characters are represented in a standard eight bit code called the ASCII code.

An example may make the difference between messages and commands a little more clear. Suppose you are trying to control a multimeter to read the resistance of a probe. The multimeter is connected to the GPIB with its address set to 21. You must first send a message to the multimeter, setting it up to read resistance (ohms). You then trigger the multimeter and read the data into the computer, using the bus. This seemingly simple operation actually involves some fairly complicated steps. First, the controller must use the Remote Enable line to bring the multimeter under control of the GPIB. The controller then sends the commands MTA0 (my talk address 0) and MLA21 (my listen address 21), indicating that the controller will be transmitting data and device 21 (the multimeter) should receive the data. At this point, the setup message is transmitted across the bus from the controller to the multimeter. The message is a set of characters that tell the multimeter to go into resistance measurement mode. An example message might be FUNCT OHMS. The message must be followed by a message terminator so that the multimeter will know that the message is complete. There will be more discussion of message terminators below.

Next, we want to trigger the multimeter. This is usually done by sending a Group Execute Trigger command, which triggers any device attached to the bus that has been previously set up as a listener.

If the computer is to read the data in from the multimeter, the multimeter must first be set up as a talker. This is done by sending the commands MTA22 and MLA0, setting the multimeter as the talker and the controller as the listener. Upon receiving these commands, the multimeter will transmit its data over the bus, once again ending the message with a terminator.

We see from the example above that the commands are used to control the activity on the bus, while the messages are the actual information that the user wants to be transmitted. Fortunately, the user rarely has to worry about the commands. Anytime that you want to send a message, the accompanying commands are required. Therefore, most software packages to control the GPIB have routines that automatically provide the appropriate commands. For example, LabView has a VI to send a message to a device. All the user has to supply are the device address and the message itself. LabView generates the necessary commands.

It is often difficult for the user to figure out the appropriate messages to send to the instrument. There are actually two different GPIB standards, IEEE-488.1-1987 and IEEE-488.2-1987. Both standards share the same hardware and command specifications, but the latter standard also specifies messages to be used for controlling common instruments.

Commands:	MTA22	Configures device 22 as the talker.
	MLA9	Configures device 9 as a listener.
	LLO	Locks the instruments panel controls.
Messages:	V+3.923	Typical data communication from a multimeter
	DCV	Message setting multimeter function to DC Voltage.

Fig. 10.7 Examples of commands and messages

Most devices do not strictly adhere to this standard. Therefore, the user must read the instrument manual carefully to determine how to control the instrument over the GPIB. A good strategy is first to learn to control the instrument's functions, using its front panel controls. Make sure you understand the operation of each function you intend to use. Then use a program that allows you to send and receive messages one at a time while you learn GPIB control of the instrument. Finally, you will be ready to develop a full program that allows you to use the instrument as part of an overall problem solution. We'll illustrate this procedure later in this chapter.

One aspect of GPIB programming that is confusing is the message terminators. The simplest way for a device to signal the end of a message is to use the EOI (end or identify) control line on the GPIB bus. However, many devices use an extra character or an extra character in combination with the EOI control line to indicate the end of the message. It is important that both the GPIB controller in the computer and the instrument use the same end-of-message terminator. In the worst case, the listener may not recognize the end-of-message terminator and may continue to wait for more characters, rather than act on the message. The two most common termination characters are LF (linefeed) and CR (carriage return), although other ASCII characters may be used. You should look in the instruments manual, which should specify the message terminator.

To avoid situations in which the controller waits indefinitely for a terminator, the LabView routines we will use have a timeout feature. Typically, the routine will stop if the end of the message has not been received within 25 seconds.

10.4 Operating the Bus from LabView

There are many different VIs in LabView to operate the GPIB, most of which you will never need. Both the tutorial and the reference manual are very confusing. However, you should not be scared off by the apparent complexity. Operating the NB-GPIB board through LabView is actually fairly simple. There are only two main functions that you will use frequently, GPIB Write and GPIB Read. These allow you to send or receive messages from an instrument attached to the GPIB. The VIs automatically produce the commands needed. Most of the other GPIB VIs in LabView are used to send commands and, as we discussed above, you will rarely, if ever, have to do this.

LabView includes two sets of VIs for controlling the GPIB, one set for the IEEE-488.2 standard and a second set called Traditional GPIB VIs. Here we will use the traditional VIs but both sets seem to work equally well.

A good place to start when working with any new GPIB device is the LabView VI called LabView<--> GPIB contained in the examples library. This VI allows you to send or receive messages from a GPIB-connected device and makes it very easy to learn about GPIB communications with a new instrument.

You will use the GPIB Read VI frequently in your GPIB programming. (See Figure 10.8.) It is used to read a message from an instrument into the computer. Address string is the address of the instrument and should be between 1 and 30. Byte count is the maximum number of characters you expect to read. LabView will stop the reading if the instrument tries to send too many characters. The mode refers to the message terminator. The default mode (0) means that the message is terminated using the EOI line and no special character. Other modes are described in the LabView manual. Sometimes you will try to read a message from an instrument that has no message to transmit. In this case the program might wait indefinitely for the message, effectively locking up the computer. The timeout is set to avoid this. The VI will wait a certain amount of time. If no message is received in that time, the VI will terminate. The timeout ms control allows two possible inputs, 0 or FFFFFFFF. If 0 is set, then timeouts are disabled. This is not a good choice for most operations. If FFFFFFFF is selected, then the default timeout is used. This is normally set to 25 seconds. You can change the default setting by running SetTimeOut, which is part of the GPIB 488.2 VIs. We usually prefer to use a timeout of a few seconds.

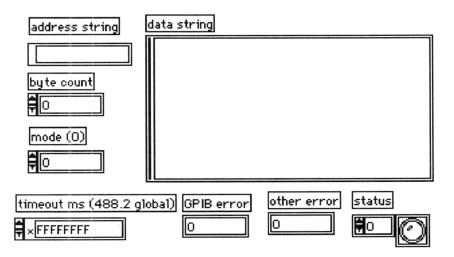

Fig. 10.8 Front panel of the GPIB Read VI

The remaining indicators are used when a problem in the message reading occurs. The most frequent problem is that no message is received. Often the problem is that the instrument needs to be triggered before it will send a message. Try using the GPIB Trigger VI before reading if you experience this problem.

The GPIB Write VI is used to send a message to the instrument indicated by the address string control input. (See Figure 10.9.) The data are put into the VI as a string of characters, which are then transmitted to the selected instrument. Mode indicates the message terminator that is used. The possible modes are described in the LabView GPIB and Serial Port VI Reference manual. You will have to look at your instrument's manual to determine which message terminator it expects. Mode 0 works for most instruments. The timeout ms input works in the same way as in GPIB Read, but it is very unlikely that a timeout will occur. A timeout usually indicates a hardware problem. Byte count tells how many characters were sent. You normally won't use it. The other two indicators are used if errors occur.

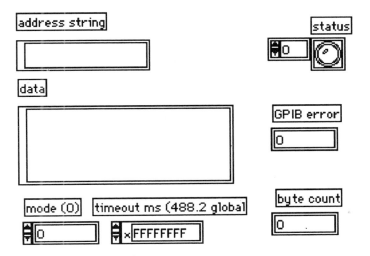

Fig. 10.9 Front panel of the GPIB Write VI

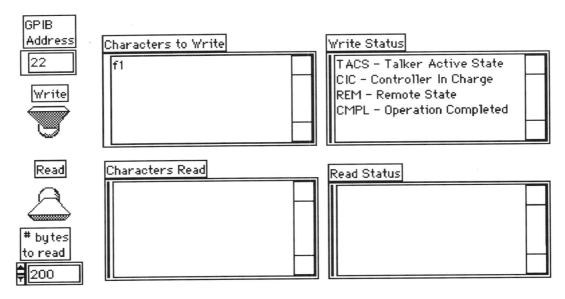

Fig. 10.10 Front panel of the LabView <--> GPIB VI

The LabVIEW<-->GPIB VI, shown in Figure 10.10, is used to communicate interactively with an instrument. To send a message, enter the instrument's address in the GPIB Address field, and type the message. Set the Write switch to true and click GO. Or you can read a message from the instrument by turning on the read switch and clicking GO. If your instrument is ready to transmit a message, it will send it over the bus. To use the read function, you must specify the maximum number of bytes to read, using the front panel control. One byte is equivalent to one character. Typically messages are from 1 to 20 characters long. The value of 200, which is the default, is a good number to use.

You should try using LabView<-->GPIB with any GPIB instrument you have available. Find the manual for the instrument, and open it to the section on GPIB programming. Try sending some messages to the instrument to see how it reacts. Many instruments have a front panel display that will allow you to determine if the message has been received properly. Try reading a message in from the instrument, too. You should be able to make the instrument operate after some practice with LabView<-->GPIB. Once you understand the programming for your instrument, you will be ready to develop a special LabView program for it.

10.5 GPIB Programming Examples

This section illustrates LabView GPIB programming for two commonly used instruments, a Fluke 8842 Multimeter and a Tektronix DM5120 Programmable Digital Multimeter. If you have either of these instruments available you should try these examples.

The Fluke 8842A is a programmable multimeter capable of measuring voltage, current, and resistance. It is programmed by sending one-, two-, or three-character messages over the GPIB A simple diagram showing the messages is illustrated in Figure 10.11. This diagram is very similar to one page of the 8842A manual, which shows in compact form all of the messages needed to program the multimeter. The arrows point to front panel buttons that perform the same function under manual control.

The Function Commands, Range Commands, and Reading Rate Commands are the most commonly used. If you have an 8842A available, try sending out a few of these

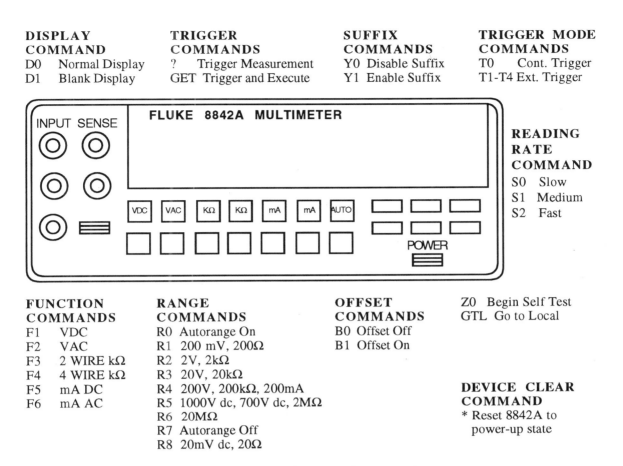

Fig. 10.11 Fluke 8842A commands

messages to the 8842A using LabVIEW<-->GPIB. You will be able to see the response of the multimeter on the display.

Reading a sample from the 8842A is simple. First, set up the multimeter for the function and range you would like. Next, turn LabVIEW<-->GPIB's Write switch off and the Read switch on; then click GO. The reading is sent over as a set of ASCII characters. You will see the message in the Characters Read window. For example, if you were to read in a voltage of 2.7462 volts, the message would be 02.7462E0.

We now develop a couple of simple examples illustrating how the GPIB VIs can be incorporated into regular LabView programs. The first example uses the Fluke 8842A multimeter. The objective is to read a set of voltage samples from the multimeter and to calculate the average. Normally, the 8842 provides voltage samples at one of three predetermined rates. For this example, we assume that the user wants to collect voltage samples at time intervals that are different from the standard intervals. We can do this by using the triggering function. We first set up the multimeter for measurement of DC voltage in a triggered mode. We then send the Trigger command over the GPIB to the multimeter. The multimeter will acquire a single voltage sample, which we may then read with the GPIB.

The front panel and block diagram for the VI are shown in Figures 10.12 and 10.13. If you have a Fluke 8842 or a compatible multimeter, we suggest you use it for your first GPIB programming. You will find it very easy to use.

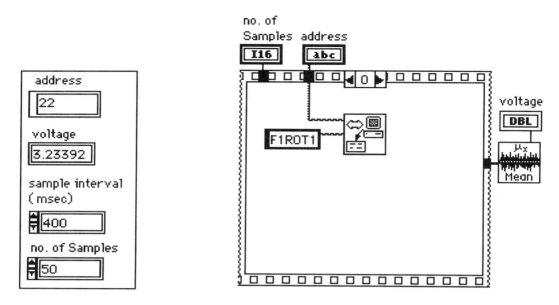

Fig. 10.12 Front panel and frame 0 of Fluke 8842 VI

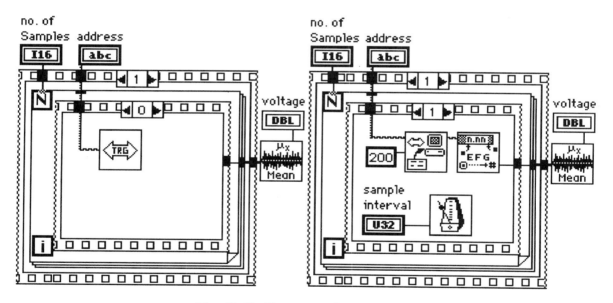

Fig. 10.13 Frame 1 and 2 of Fluke VI

The front panel includes controls for the instrument address, the sample interval in milliseconds, and the number of samples to include in the average. The first frame of the diagram is just GPIB Write used to send a string of characters to select measurement of DC voltage (F1) with autoranging (R0) and triggered operation (T1).

Frame 1 of the main sequence contains a For Loop, which is executed once for each sample. Within the loop is a second sequence structure. The trigger signal is sent out in frame 0 of the internal sequence. The trigger VI is very simple, requiring only the instrument address as an input.

The second frame of the inner sequence includes GPIB Read, which is used to read the voltage sample from the instrument. The constant 200 connected to it is the byte count. Note that we have used the default inputs for everything but the address. Each voltage sample is actually represented as a string of characters. These strings must be converted to number formats before we can process the data. The string conversion function in the upper right converts a character string in the E format used by the multimeter into an actual number. The Wait Until Next Millisecond Multiple function causes a set delay each time the loop executes. The samples pass out of the loop as an array and the average is computed using the Mean VI.

The Tektronix DM5120 is a standard 6 1/2-digit multimeter that also incorporates some more advanced features. These include the capability to make high-speed readings (1,000 samples/sec) at reduced measurement resolution, the ability to subtract automatically a preprogrammed "null" value from each measurement, on-board storage for up to 500 readings, a programmable digital filter to smooth noisy signals, and the ability to send text messages to the multimeter display. All of the standard and advanced features can be programmed over the GPIB.

Programming the DM5120 is more complicated than programming the Fluke 8842. Approximately 30 pages of the manual are dedicated to GPIB programming of the multimeter. Some of the standard functions are quite easy to program. For example, the standard voltage, current, and resistance functions are selected by sending simple codes like DCV for DC voltage and OHMS for resistance measurement. Once you get used to programming these standard features over the GPIB, it is not too difficult to follow the manual and to learn to program the more complex features.

There is one aspect of the programming that may cause confusion. With some multimeters you can take readings repeatedly without triggering. For example, with the Fluke 8842, you can use the GPIB Read VI to read successive voltage samples. However, when programming the DM5120 you must either trigger the multimeter or write a message to it between each read.

The following example will illustrate some of the programming features of the DM5120. The example is for a situation in which the resistance of a conductor is being measured during a heating process. The conductor will be heated continuously until it melts at which point the connection will be broken and the resistance will go to a large value. The DM5120 is used to measure the resistance repeatedly. When the resistance exceeds some threshold, a message is sent to the multimeter display, telling the experimenter that the sample has melted.

The front panel of the DM5120 VI is shown in Figure 10.14. This VI is designed to measure resistance values continuously until the measured resistance exceeds the Comparison Value. When this happens, a user-selected message will be displayed on the multimeter display. The message can be up to ten characters long, which is the full width of the DM5120 display. The front panel includes controls for the instrument address, the sample interval in milliseconds, and the resistance comparison value.

Frame 0 shows GPIB Write used to send a string of characters to select measurement of resistance (OHMS) with autoranging (Range 0). Frame 1 contains a While Loop with another Sequence Structure inside it. Frame 0 within the While Loop contains the Trigger function.

In frame 1 within the While Loop (Figure 10.15), GPIB Read is used to read the resistance sample from the instrument. The constant 200 connected to it is the byte count. Note that we have used the default inputs for everything else except the address. Each resistance sample is actually represented as a string of characters. These strings must be converted to number formats before we can process the data. The string conversion function in the upper right converts a character string in the E format used by the multimeter into an actual number. The Wait Until Next Millisecond Multiple function causes a set delay each time the loop executes. When the resistance value is greater than the comparison value, the While Loop stops.

104 The IEEE-488 General Purpose Interface Bus

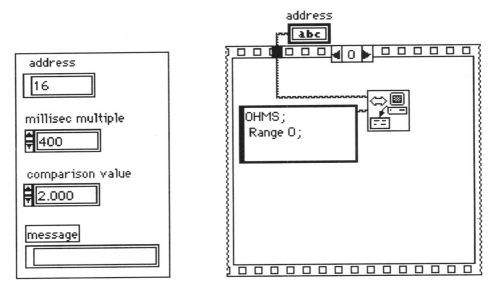

Fig. 10.14   Front panel and frame 0 of DM5120 VI

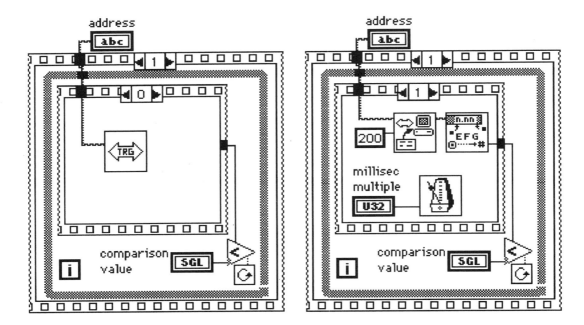

Fig. 10.15   Frame 1 of DM5120 VI

Frame 2 of the external sequence structure (Figure 10.16) writes the completion message onto the display of the DM5120. Characters are written on the display by sending the message text "characters" to the DM5120 over the GPIB. The user of this VI will enter a string that contains only the actual characters to appear on the display. Here we use the Concatenate Strings function to build the appropriate string to send to the DM5120. The string sent over the GPIB is text" followed by the user-specified characters followed by a closing quotation mark (").

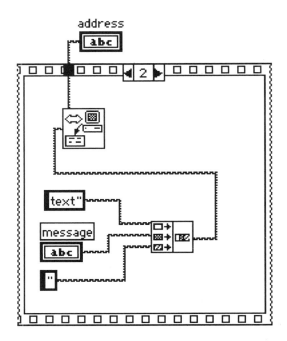

Fig. 10.16 Frame 2 of DM5120 VI

National Instruments has developed a library of virtual instruments to control many popular instruments using the GPIB. These can be obtained directly from National Instruments after you purchase LabView. You can use these directly or just use them as programming examples for development of your own VIs. In many instances, the instrument control VIs are quite complicated, since they must be able to control all of the different instrument functions. You may not need many of these functions in your own application. We recommend that you try to program your own GPIB interface first, since in many instances it will be quite simple.

Chapter 11

Analysis of Sampled Data

11.1 Introduction

We frequently use our computer to examine time-varying signals from external transducers. The time dependence can come from a variety of sources. In many cases, the signal is nominally steady but may be contaminated by electronic noise. For example, the voltage output from a temperature sensor may fluctuate even though the temperature is constant. In such a case, you would use the computer to acquire a set of samples of the voltage output and compute the mean voltage reading. You then need statistical tools to estimate the uncertainty in the resulting mean temperature reading. Other times the signal may be periodic, as in the case of a microphone sensing acoustic waves from a whistle. In this case, you may desire to compute the power in the signal or extract the waveform. A time-varying signal may also arise from a random process such as an air velocity sensor in a turbulent flow field. Finally, the signal may be transient, such as a force sensor measuring a rocket's thrust during startup. Here the challenge may be to extract the actual transient behavior from a noisy transducer output.

In this section we will first discuss sampling of a signal, then provide formal definitions and a classification of random variables. We will focus initially on statistical descriptors of a stationary random variable, including the probability density function, the mean and standard deviation, and higher moments of the probability distribution. In each case we will give the general definition, discuss rules for selecting the sample size, develop appropriate LabView instruments, and, where appropriate, give a practical example of the application. Continuing with stationary random variables, we will then examine statistics that represent the time variation of the signal, including the power spectrum and the autocorrelation. We will also describe the cross-correlation between two different signals. The section will conclude with a description of special techniques for examining periodic (or nearly periodic) signals and transient processes. This tutorial will cover only the most commonly used techniques for processing the sampled data record. There are many advanced techniques in digital signal processing that are well beyond the scope of this tutorial. Some of these techniques are supported by LabView so are within the reach of everyday users.

Data Sampling

It is important to understand clearly the concept of sampling before discussing the applicable analytical techniques. Usually, a physical quantity of interest varies continuously. However, when we use a transducer and a computer to record the quantity, we can do so only at a finite number of discrete times. A typical example is the use of an A-to-D converter to acquire a sequence of voltage measurements from a microphone, as illustrated in Figure 11.1. Here the continuous line represents the microphone output, and the dots represent the voltage samples.

It is important to realize that the computer cannot record the entire continuous signal; it can only acquire samples. The samples may be spaced closely enough that they allow an excellent reconstruction of the continuous signal. However, in other cases the samples

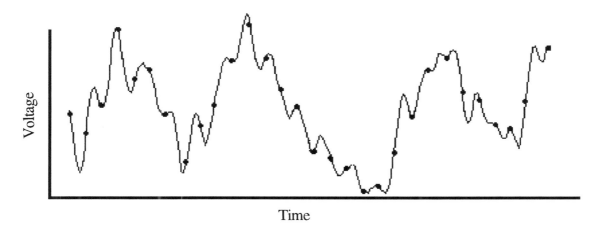

Fig. 11.1 The continuous signal from a microphone and discrete samples at regular time intervals

may be very widely spaced in time, providing only an occasional sample of the level of the signal.

We call the set of samples a sampled data record. Statisticians might describe our sampling process as taking a sample of a population. This is not a perfect description since a population refers to a finite number of discrete elements. In this case we are sampling a continuous signal that can be infinitely divided. Nevertheless, we can use the statistician's tools. Frequently in laboratory work we desire only a statistical description of the signal, finding, for example, the standard deviation or the power spectrum. We then use statistical tools to estimate how well our samples represent the true signal.

The sketch above represents periodic sampling of a continuous variable. This means that successive samples are acquired at a constant time intervals. This is the most common way of sampling a signal. In many cases a transducer produces a continuous analog voltage, and a periodically sampled data record is produced, using an analog-to-digital converter and a routine such as AI Acquire Waveform. Periodic sampling is required for some of the analytical techniques, such as the use of the Fourier transform techniques to estimate the power spectrum. However, many of the analytical techniques do not require periodic sampling, and it is used only for convenience.

Some transducers provide samples only at random times, even if the underlying physical process provides a continuous signal. An example is a laser-Doppler velocimeter, which measures fluid velocity by measuring the velocity of small tracer particles. Velocity samples are available only when randomly arriving particles are present in the measurement volume. Such data may also be processed statistically, but special care is sometimes required.

An important consideration in planning a data acquisition program for time-dependent data is the selection of the data sampling rate. Returning to the example above, we see that the samples are too widely spaced to reconstruct the continuous signal. If we wanted to reproduce precisely the sound that produced the original signal, we would have to sample at a somewhat higher rate. However, in most scientific or engineering work we are interested only in the statistical descriptors of the signal, so an exact representation of the full continuous signal is not necessary.

The selection of the sampling rate is highly dependent on the type of analysis that will be applied to the sampled data record. If you wish to estimate the mean and standard deviation of the signal, you should sample the signal slowly. The individual samples in the record must be statistically independent in order to apply standard analytical tools for calculating the uncertainty in the estimates. If the sampling rate is too high, estimates of the mean calculated from separate data records will not agree.

As an illustration of the problem of sampling too fast, we show in Figure 11.2 a four second record of a continuous voltage signal. Imagine that a set of 500 samples was acquired at a rate of 1,000 samples per second. The total sampling time would be 0.5 seconds, as illustrated. Clearly, the mean voltage calculated from the 500 samples would not accurately represent the true mean level of the signal. Acquiring the same number of samples more slowly would allow a more accurate estimate of the mean. On the other hand, if we are interested in estimating the power spectrum of a signal, we must use a high sampling rate. The rule of thumb is that the sampling frequency must be at least twice the highest frequency in the signal. The sampling rate criteria for power spectrum estimation will be discussed in detail later in this chapter.

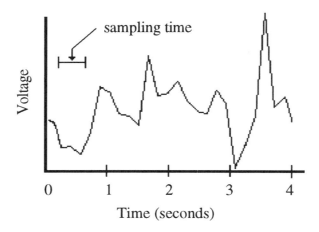

Fig. 11.2 An example of how sampling too fast produces erroneous results

11.2 Random Variables

In this chapter we are developing analytical techniques that we will apply most often to random variables. In some cases, the quantities that we measure may vary randomly because the underlying physical process has a random nature. In that case we may be using our analytical techniques to characterize the stochastic nature of the physical process. Alternatively, the physical process we are studying may provide a stable signal, but our measurements are subject to random errors. In this case we may be trying to estimate the true value of the signal, and our confidence in our measurement of that value.

Before beginning our discussion of the analytical techniques we will make some formal definitions and discuss the relation between time averaging and ensemble averaging. We will then give a brief discussion of measurement techniques for nonstationary processes. Finally, we will limit our discussion to stationary random variables for the bulk of the analysis chapter.

A random variable is one in which the time variation of the variable cannot be described by an explicit mathematical function. One cannot predict the value of the variable "very far" in the future, even knowing the value at the present time and all past times. How far in the future one can predict the value depends on the specific problem. For example, if the random variable were the air temperature outside your window, you could accurately predict what the temperature would be one second or even one minute in the future. However, you could not predict the temperature 24 hours later with any certainty.

The opposite of random is deterministic, the most common deterministic variables being sine waves or other periodic functions. No physical system will produce a perfectly periodic signal, since a transducer signal will always be contaminated by some level of noise. However, in many situations the noise level is insignificant. An example is the signal from an accelerometer attached to a lathe rotating at constant speed. The signal is dominated by periodic accelerations related to imbalances in the lathe spindle and contaminated by electronic noise and random vibrations of the system. Unless the lathe is very precisely balanced, the periodic signal will dominate.

Most of our analytical methods will be derived for random variables that are continuous functions of time, as illustrated in Figure 11.3. Note, though, that when the data are acquired by the computer, they will be sampled only at discrete times. The single time history of the random variable recorded for a finite time is called a sample record. Conceptually, we may think of the sample record as part of a time history of infinite duration, called a sample function. The random process is then defined as the collection of all possible sample functions. Practically, we may acquire either a single sample record or a set, or ensemble, of sample records. If the process is truly random, each sample record in the ensemble will differ from all others.

To make the definitions clear, we consider an experiment in which the noise in the signal from a disk-drive read head is under investigation. The magnetic read head produces a voltage signal that nominally contains only two signal levels representing either a 1 or a 0. However, the signal is contaminated by considerable noise that arises from the electronics, from the strong magnetic fields present, and from vibrations of the read head induced mechanically or by air turbulence in the disk drive. The experiment may be specified as: i) write a track containing all 1's. ii) begin sampling the head output when at a reference position. iii) continue to acquire data for ten revolutions of the disk. The experimental setup and procedure are then the random process. A single time series of the read-head signal is a sample record that, if it continued to infinite time, would be called a sample function. Usually we would expect to repeat the experiment a number of times, producing an ensemble of sample records.

We are now prepared to discuss averaging and classification of random data. We consider two types of averaging, time averaging and ensemble averaging. The time average is defined on a single sample record as:

$$\mu_x = \frac{1}{T} \int_0^T x(t) dt.$$

The ensemble average is defined over an ensemble of sample records as:

$$\mu_x(t_0) = \frac{1}{N} \sum_{i=1}^{N} x_i(t_0).$$

Fig. 11.3 A single sample record from a random variable

110 Analysis of Sampled Data

This is illustrated in Figure 11.4 for an ensemble of four records. Note that the values from all the records at the specific time are averaged. Time and ensemble averaging are not necessarily interchangeable, as will be discussed below.

A random process is classified as stationary if ensemble averages performed over a large number of sample records are independent of the time at which the samples are taken. If the process illustrated in the last example were stationary, we would be able to chose another time t to perform the averaging and would expect to get the identical result. Of course, we would have to use a much larger set of sample records if we expected to obtain precise agreement.

A stationary random process may be further classified as ergodic or non-ergodic. *Ergodic* means that time averages computed over different sample records from the same random process are the same. It follows, then, that for an ergodic random process, time averages over a single sample record may be used in place of ensemble averages over multiple records. Fortunately, most stationary random variables arising from physical processes are ergodic, so we can use a single sample record to compute the statistical descriptors of the process.

A nonstationary random process is one in which ensemble averages are a function of the time chosen for sampling. One example is the thrust of a rocket motor during startup. The plots in Figure 11.5 are measurements of the thrust. Each plot starts at time equal to zero when the fuel valves are just opened and the thrust is zero. The thrust increases, asymptotically approaching a nominally constant thrust after some time. The ensemble average measured over a large number of rocket firings shows this nonstationary trend clearly. The single sample record shows that the process is truly random. The overall transient behavior is still obvious, but a substantial fluctuation is superimposed. The only way to determine the average behavior of the thruster is to ensemble-average. A time average of the single sample record would be meaningless in this case.

For a more subtle example of a nonstationary random process, we return to our example of the noisy signal from a disk-drive read head. Recall that we stated the process as: start sampling the signal at a reference position on the disk. This is a nonstationary process. Consider that the disk must be at least slightly wavy. This is likely to cause noise that is phase-locked to the read head's position over the disk. Therefore, we would not expect the ensemble average for a time when the head is aligned with a peak of the waviness to be the same as the average for a time when the head is aligned with a valley.

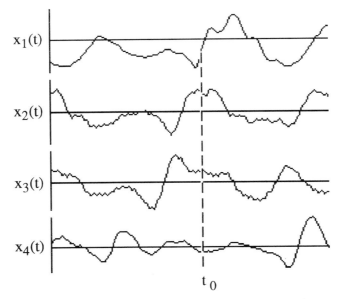

Fig. 11.4 An ensemble of four sample records for the disk drive noise test

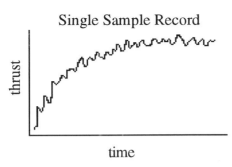
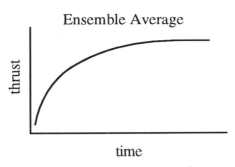

Fig. 11.5 A single sample record and the ensemble average of the rocket motor thrust during start-up

This example is subtle because the process would be stationary if we were to change the experimental description slightly. If we instead started acquiring each sample record at a random time, the waviness would be uncorrelated with the sampling. That is, it would be equally likely that a given sample of the record would be aligned with a peak as with a valley. Therefore, the noise induced by the waviness would appear simply as an additional stationary random noise.

11.3 The Probability Density Function

The Probability Density Function (PDF) of a sample record describes the probability that a random variable will be in a certain range. Formally, the probability density function is defined as:

$$p(x) = \lim_{\Delta x \to 0} \frac{\text{Prob }[x \leq x(t) \leq x+\Delta x]}{\Delta x}.$$

This means that the probability that the variable x is within a small interval Δx of x_0 is equal to $p(x_0)\Delta x$. The probability that the random variable is between two values x_1 and x_2 at some instant is:

$$\text{Prob }[x_1 \leq x(t) \leq x_2] = \int_{x_1}^{x_2} p(x)dx.$$

The probability that the random variable will have a value between negative and positive infinity is, of course, 1 so:

$$\int_{-\infty}^{\infty} p(x)dx = 1.$$

The PDF may be thought of as the limiting case of a histogram. To make a histogram, the full range of a random variable is divided into even intervals. As the variable is sampled, the sample is attributed to one of the intervals. This is often referred to as "binning" the samples. You can imagine throwing each sample into the appropriate bin as it comes in.

An example of a histogram is shown in Figure 11.6 for a voltage signal that varies between 0 and 5 volts. The voltage range has been divided into 10 even intervals and a total of 50 samples divided among the intervals. A much larger number of intervals and samples are needed if the histogram is to form a good representation of the PDF of the

112 Analysis of Sampled Data

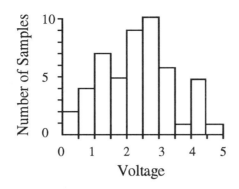

Fig. 11.6 A histogram of voltage samples

random variable. In practice we approximate the PDF using a histogram. LabView has a built-in utility to compute a histogram from a one-dimensional array. We demonstrate below how to use this utility and how to normalize the resulting histogram.

We have developed a VI, called PDF, that acquires a set of voltage samples through the A-to-D converter and plots the data in a normalized histogram form. The front panel shown in Figure 11.7 is similar to AI Waveform Scan, with the addition of a control to specify the number of intervals to be used in the histogram. The plot is an XY Graph. The histogram values are normalized by the product of the number of samples and the interval size so that the integral under the curve is 1.0.

The diagram for PDF is shown in Figure 11.8. The upper half of the diagram is AI Waveform Scan connected to all of its inputs from the front panel. The output array is converted to a one-dimensional array using the Index Array function. The bottom half shows the data put into a histogram form. The result is normalized by dividing each value in the histogram by the histogram's area. The area of the histogram is calculated by multiplying the number of samples by the histogram interval size. The interval size is calculated by subtracting the second interval value (x value) from the first interval value. The x value is determined using the Index Array function. The x and y values are put into a Bundle function to build a cluster. The normalized histogram is plotted on an XY Graph.

Generally, many samples are required to obtain a histogram that is a reasonably accurate representation of the PDF. Shown in Figure 11.9 are four histograms with 50 bins, each formed from the same data set using 200, 500, 1,000, or 2,000 samples. Even with 2,000 samples, the histogram is still not very smooth.

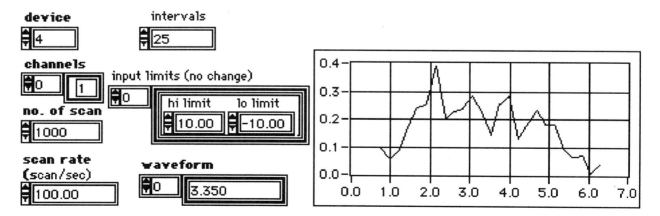

Fig. 11.7 Front panel of the PDF VI

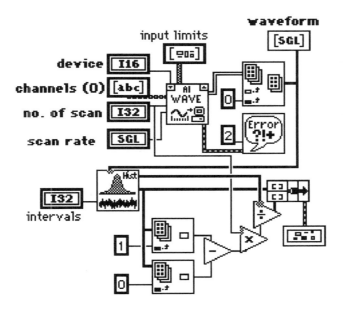

Fig. 11.8 The diagram for the PDF VI

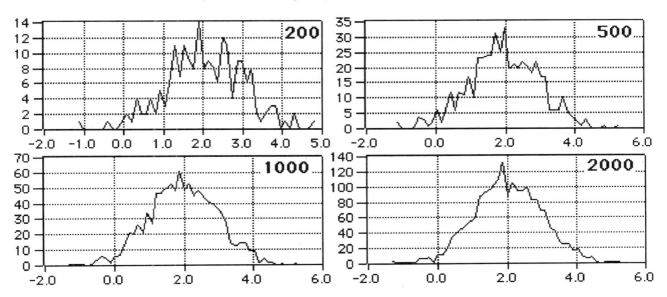

Fig. 11.9 Four PDFs from the same sample data record using different numbers of samples

The plots above were made using a test data set generated by a Gaussian random number generator. The test data sets, called TestDat1 and TestDat2, are available on the VI disk. You may want to use them to test your various routines. To read the data, you can use the standard file handling techniques, used, for example, in the PlotFile VI. The sample size is 2,000 for TestDat1 and 5,000 for TestDat2. The mean value for each data set is 2, and the standard deviation is 1. Figure 11.10 shows the panel and the diagram of a VI to read the data set and plot a histogram. The data sets can be used equally well to test other statistical processing routines. Keep them in mind as you try out the various VIs.

114 Analysis of Sampled Data

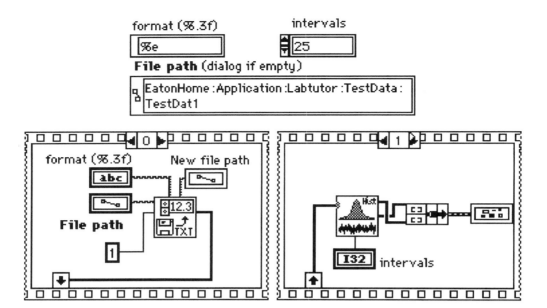

Fig. 11.10 The panel controls and diagram of a VI to read one of the test data sets

The Gaussian or Normal Probability Density Function

The Gaussian, or normal, distribution plotted in Figure 11.11 is defined as:

$$p(x) = \frac{1}{\sqrt{2\pi}\,\sigma} \exp\left[-\frac{1}{2}\left(\frac{x-\mu}{\sigma}\right)^2\right]$$

where μ is the mean value and σ is the standard deviation. This is the well known "bell-shaped" curve that you may have experienced in previous studies of experimental uncertainty. The Gaussian distribution is important because it is generally an excellent approximation of the distribution of random errors that occur when a fixed value is measured repeatedly. For example, if you use a force transducer connected to a voltmeter to measure the weight of an object, you would expect the set of measurements to fill a Gaussian distribution centered on the actual weight of the object. This follows from the Central Limit Theorem discussed below. Many other physical processes also produce random variables with nearly Gaussian distributions.

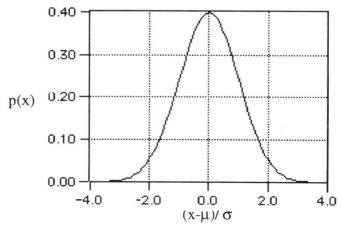

Fig. 11.11 The Gaussian probability density function

Consider a random variable that is itself a sum of a large number of independent random variables.

$$w(t) = \sum_{i=1}^{N} u_i(t).$$

Each of the random variables $u_i(t)$ has the same distribution; that is, the shape of the PDF is the same, but the mean and standard deviation need not be the same. The Central Limit Theorem then states that the random variable w(t) has the Gaussian distribution regardless of the underlying distribution of the component random variables. For example, each of the u_i's may be uniformly distributed over an interval, which means that any value in the interval is equally likely to occur. The sum w(t) is still normally distributed. We do not prove the Central Limit Theorem here, but a proof appears in many texts on probability theory.

The importance of this is that measurement errors are generally due to the sum of many different possible error sources. If the error sources are truly independent (which is often the case), then a set of measurements will be normally distributed. The assumptions needed for the Central Limit Theorem appear quite restrictive, but the result is actually broadly applicable.

Examples of the Application of the Probability Density Function

Often we use the probability density function in the early phases of an experiment to check the performance of our measurement system. We usually expect continuous systems to produce a nearly Gaussian PDF. If we get strong deviations, we might suspect a problem with our measurement system. The plot in Figure 11.12 is from the PDF VI connected to a transducer that can produce occasional incorrect signals if set up improperly. The main peak of the distribution represents the true variation of the variable. However, the smaller peak to the right of the main peak represents noise signals. If the full sampled data record were used to calculate the mean value, the result would be wrong. The PDF has warned us of a problem that we may be able to correct by adjusting the transducer. Alternatively, we may be able to use the PDF to estimate the true statistics of the signal.

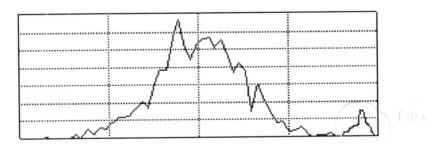

Fig. 11.12 Measured PDF showing an extra peak on the right side due to transducer problems

A second example of a pathological PDF is illustrated in Figure 11.13. This histogram was recorded from data transmitted digitally from the transducer to the computer. The main body of the PDF appears nearly Gaussian, but there is a narrow peak surrounded by dips to the left of the main peak. This suggests a problem with the digital transmission. It may be that one bit in a parallel digital connection is faulty. This would cause certain values to be transmitted incorrectly. A PDF such as the one shown here will result if a high-order bit is faulty. If a lower-order bit is the problem, there will probably be multiple spikes in the

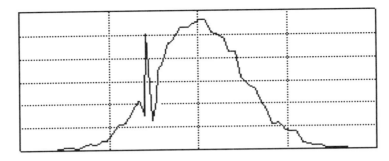

Fig. 11.13 PDF showing a spike surrounded by dips characteristic of a faulty digital data transmission

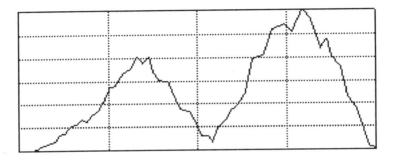

Fig. 11.14 Example of a bimodal PDF

PDF. Again, the PDF has provided us with a diagnostic for problems in our measurement system.

Figure 11.14 is an example of a non-Gaussian PDF that is a not a result of instrument error. This is called a bimodal PDF because it appears that the random variable fluctuates around two distinct states. Such a PDF occurs in some physical systems when there are two quasi-stable states. The variable may fluctuate around one of the states for a time, then suddenly shift to fluctuating about the other state. This example illustrates how the PDF may be used to learn about the physical behavior of a system.

Exercise 1

Use your function generator and the PDF VI to examine the probability density functions of a sine wave, a square wave, and a triangle wave. They should look like the PDFs in Figure 11.15. You have to be a little careful in setting up the sampling rate. If the sampling rate is an integer multiple of the function generator frequency or, worse yet, a rational fraction of the frequency, the PDF will look strange. It is probably best to set it up so that the sampling rate is considerably faster than the wave frequency.

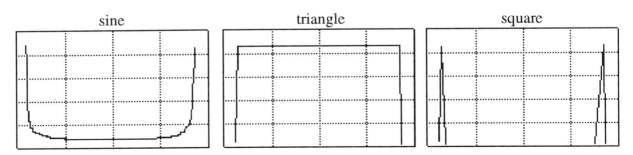

Fig. 11.15 PDF plots for the sine, triangle, and square waves

11.4 Mean and Standard Deviation

The most commonly used scalar descriptors of the probability density function are the mean and the standard deviation. In many instances the mean value is all that is desired, as when all the random variation in a signal is just associated with noise. In other cases, we wish to know the power associated with oscillations or random fluctuations of a signal, so we require a measure of the standard deviation. The mean value of a random variable is defined as:

$$\mu_x = \lim_{T \to \infty} \frac{1}{T} \int_0^T x(t) dt.$$

Of course, we will almost always estimate the mean from a set of discrete samples, in which case we define the sample mean as:

$$\bar{x} = \frac{1}{N} \sum_{i=1}^{N} x_i.$$

where x_i indicates the ith sample and the symbol $\bar{x}$ is used for the sample mean to indicate that it is an approximation of the true mean.

The variance, which is the square of the standard deviation is defined as:

$$\sigma_x^2 = \lim_{T \to \infty} \frac{1}{T} \int_0^T [x(t) - \mu_x]^2 dt.$$

where σ_x is the standard deviation. The standard deviation is a measure of how much the signal deviates from the mean value. It is also often called the root mean square value, or rms.

Usually, we will be calculating an estimator for the variance from a set of samples. The sample variance is defined as:

$$s^2 = \frac{1}{N-1} \sum_{i=1}^{N} (x_i - \bar{x})^2.$$

You may be surprised to find that the denominator is N-1 rather than N. Statistics textbooks demonstrate that if the denominator is N, the estimator is biased. In practice, we normally acquire fairly large numbers of samples with our computer, so the difference between N and N-1 is insignificant.

The LabView Mean and Standard Deviation VI

Calculating the mean and standard deviation of an array of samples is simple in LabView. The diagram in Figure 11.16 is from the Vave VI. The AI Acquire Waveform VI acquires a set of voltage samples from the A-to-D converter. The subVI at the right then computes the mean and standard deviation. This subVI is called Standard Deviation and is available under the Statistics submenu of the Analysis Function menu.

118   Analysis of Sampled Data

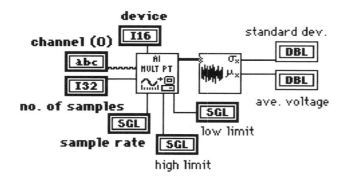

Fig. 11.16 A diagram showing the use of the LabView mean and standard deviation VI

Calculations With Accumulated Sums

Sometimes you want to compute the mean and standard deviation without storing an array of samples. The values can instead be calculated using only the sum of the sample values and the sum of the squares of the sample values, as shown below. Thus, only two numbers need be stored, rather than a whole array of samples. To see how we can calculate the sample variance, we start with the definition and multiply out the squared term:

$$s^2 = \frac{1}{N-1} \sum_{i=1}^{N} (x_i - \bar{x})^2 = \frac{1}{N-1} \sum_{i=1}^{N} (x_i^2 - 2x_i\bar{x} + \bar{x}^2)$$

$$= \frac{1}{N-1} \left[\sum_{i=1}^{N} x_i^2 - \sum_{i=1}^{N} 2x_i\bar{x} + \sum_{i=1}^{N} \bar{x}^2 \right].$$

We see that the last term is just N times the mean value squared and the first term is just the sum of the square of the samples. Recall that:

$$\sum_{i=1}^{N} x_i = N\bar{x}.$$

The variance may then be calculated as:

$$s^2 = \frac{1}{N-1} \left[\sum_{i=1}^{N} x_i^2 - 2N\bar{x}^2 + N\bar{x}^2 \right] = \frac{1}{N-1} \left[\sum_{i=1}^{N} x_i^2 - N\bar{x}^2 \right].$$

We calculate $\bar{x}$ directly from the sum of the samples, then we can calculate the variance and the standard deviation using the sum of the squares of the sample. We use this simple trick to calculate running averages quickly in the following LabView example.

Rather than first acquiring an entire array before computing statistics, we may wish to display a running mean and standard deviation computed continuously as we sample the data. To do this we use the formulas developed previously to compute the mean and the standard deviation from the sum of the samples and the sum of the squares of the samples. The panel for such an instrument, called RunAve, is shown in Figure 11.17. The user must specify the device number, channel, and high and low limits. The sample size is not specified. When you run the VI it will continue to sample the data and recompute the mean and the standard deviation after each sample. The present values are shown along with the total number of samples. To stop the sampling, you just click on the stop sign. Try running the VI using a function generator as the input.

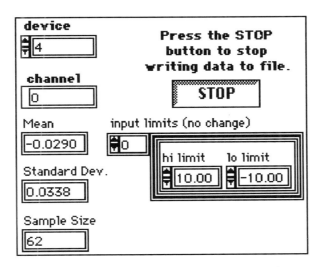

Fig. 11.17 Front panel for the RunAve VI

The diagram for calculating the running mean and standard deviation is shown in Figure 11.18. The mean and the standard deviation are calculated with each cycle of the While Loop. AI Read One Scan acquires a single voltage sample with each cycle of the loop. The output of AI Read One Scan is converted to an element using the Index function. The mean and the standard deviation are calculated in the formula node. Xm is the running mean. Xi is the sample voltage. Xb is initially zero, and the sum of the voltage samples thereafter. Sd is the standard deviation. Xs is initially zero and the sum of the voltage samples squared thereafter. The sample size is displayed as i plus one, since i is initially zero.

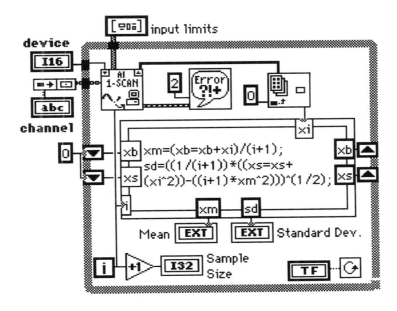

Fig. 11.18 Diagram for the RunAve VI

Estimates of Statistical Uncertainty

Estimates of the mean and the standard deviation will deviate from the true value because of the finite sample size. For example, if we collect five separate sample records of a stationary random variable, the sample mean computed from each record will be different from all the others. We need to use statistical tools to estimate the range of possible deviation from the true value.

The sample mean is itself a normally distributed random variable. That is, if we acquire a large number of sample records from the same stationary random variable, the sample means will fill a normal distribution regardless of the probability density function of the underlying random variable. This follows from the Central Limit Theorem if we assume that the N samples forming an individual sample record are statistically independent. The mean of the distribution is just equal to the mean of the original random variable. The standard deviation depends on the sample size. It is intuitive that the larger the number of samples used to compute the sample mean, the smaller the deviation from the true mean will be. In fact, the standard deviation of the distribution decreases as the reciprocal of the square root of the number of samples.

The theoretical (normal) distribution for the sample mean is shown in Figure 11.19. As shown, the mean value of the distribution is just equal to the mean of the random variable x. This is true because the sample mean is an unbiased estimator of the mean. The standard deviation of the distribution is the standard deviation of the random variable x divided by the square root of the number of samples. This is very intuitive. If the standard deviation of the original random variable is small, then the mean can be accurately estimated with only a few samples. Alternatively, if the number of samples in each sample record is small, then we expect a large variation between the sample means of the different sample records.

Each of the three histograms in Figure 11.20 shows the distribution of sample means for 200 sample records. That is, 200 sample records were collected, the mean calculated for each sample record, then the ensemble of means plotted as a histogram. The original random variable came from a random number generator. For the left plot, each sample record contained only 100 samples, and the distribution of sample means is quite wide. For the middle plot, each sample record contained 400 samples, and the distribution is about half as wide. For the plot on the right, each sample record contained 1,600 samples, so the distribution is about half as wide as in the middle plot. This demonstrates empirically what was discussed intuitively above.

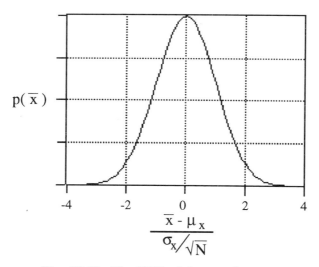

Fig. 11.19 The PDF of the sample mean

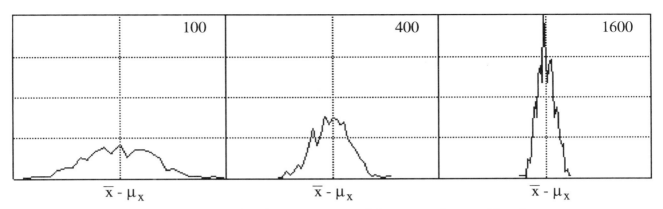

Fig. 11.20 Histograms of the sample mean for three different sample record lengths

Normally, when we measure the mean value of a random variable, we acquire only a single sample record. What we need is a method to estimate how far the sample mean from this single sample record might deviate from the true mean. This brings us to the concept of a confidence interval. We would like to state that the true mean is within a certain interval around the sample mean at a given confidence level. For example, we might write:

$$\bar{x} - \Delta x < \mu_x < \bar{x} + \Delta x \text{ at } 95\% \text{ confidence}$$

The 95 percent confidence level implies that there is a 5 percent chance that the true mean is outside of this interval. Since we know that the sample mean is a normally distributed random variable, we can determine the probability that the random variable is in any interval. For example, the probability that the deviation is less than $2\sigma_x/\sqrt{N}$ is the integral of the normal distribution from -2 to 2, as illustrated in Figure 11.21.

$$\text{prob}\left[\left|\bar{x} - \mu_x\right| < \frac{2\sigma_x}{\sqrt{N}}\right] = \frac{1}{\sqrt{2\pi}} \int_{-2}^{2} e^{-x^2/2} \, dx = 0.954.$$

We then say that the deviation of the sample mean from the true mean is less than $2\sigma_x/\sqrt{N}$ with 95.4 percent confidence. The integral was evaluated using standard normal distribution tables in a statistics book. We can use the same tables to find the confidence interval for other confidence levels. A short table is given in Figure 11.22 for confidence levels that you may want to use. For example, the confidence interval Δx is $1.96\sigma_x/\sqrt{N}$ at the 95 percent confidence level. This means that we have 95 percent confidence that the

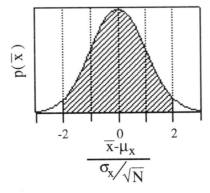

Fig. 11.21 The integral over the hatched area is the probability that the mean is within the interval

Confidence Level	Confidence Interval
0.50	.675 $\sigma_x/\sqrt{N}$
0.90	1.65 $\sigma_x/\sqrt{N}$
0.95	1.96 $\sigma_x/\sqrt{N}$
0.99	2.58 $\sigma_x/\sqrt{N}$

Fig. 11.22 Table of confidence intervals for various confidence levels

sample mean is within Δx of the true mean of the signal. You can, of course, use any confidence interval you would like. The 95 percent confidence level is used most frequently in standard science and engineering practice. However, higher confidence intervals may be used if the consequences of being wrong are severe. For example, in doing experiments in support of nuclear power plant design a high confidence level would be appropriate.

Consider an example in which a sample record of 1,000 samples was acquired. The sample mean is 4.58, and the sample standard deviation is 1.23. Assume that you want to calculate the statistical uncertainty in the sample mean at the 95 percent confidence level. The confidence interval is:

$$\Delta x = 1.96 \, \sigma_x/\sqrt{1000}$$

We have no option but to use the sample standard deviation, since we do not know the true standard deviation:

$$\Delta x = 1.96 \times 1.23/\sqrt{1000} = 0.076$$

Then we can say that the true mean of the random variable is in the range:

$$4.58 - 0.076 < \mu_x < 4.58 + 0.076, \quad 4.50 < \mu_x < 4.66 \text{ at } 95\% \text{ confidence.}$$

If the sample record had contained 2,000 samples, the confidence interval would be smaller:

$$\Delta x = 1.96 \times 1.23/\sqrt{2000} = 0.054, \quad 4.53 < \mu_x < 4.63 \text{ at } 95\% \text{ confidence.}$$

Our analysis of confidence intervals has assumed that the samples in the sample record are statistically independent. However, if the sampling rate is too fast, the samples will not be independent. This will have the effect of reducing the number of samples and broadening the confidence interval. This provides an easy and important check of your sampling rate. An appropriate procedure is:
1. Acquire several sample records all of the same length.
2. Calculate the sample mean and the confidence interval for each record.
3. Check that the sample means all agree within the estimated confidence intervals.
4. If they do not agree then you are probably sampling too fast. Decrease your sampling rate and repeat the procedure.

A good rule to remember when acquiring data to estimate the mean and the standard deviation is: When in doubt, sample more slowly! There is frequently confusion on this point, so you should make sure you understand the reasons for this rule.

The MeanComp VI

Rather than collecting a certain length sample record and then calculating the confidence interval, we may want just to check that the measured mean value repeats to within a given percentage. The MeanComp VI performs such a function. Initially, MeanComp collects two equal length sample records, then calculates and compares the means. If the difference in the means is smaller than a specified percentage, the mean of the two groups combined is displayed. If the difference is larger than the specified percentage, a third group of data, with sample size equal to the first two groups combined, is collected. The mean of the third record is compared to the mean of the first two groups combined. If the difference in means is still larger than the specified percentage, the third group is combined with the first two groups and a new mean is calculated. This process is continued until the means are within the specified percentage. The sizes of the two groups being compared are always equal, and the final mean value displayed is computed from the entire sample.

The MeanComp VI is described here for completeness. However, understanding of this VI is not needed to understand the remainder of this section, and you may skip this part of the book.

The front panel for MeanComp (See Figure 11.23) is the same as VseqP2, with the addition of the specification of the percentage difference in mean. The initial sample size is the amount of data collected for each of the first two groups. The mean displayed is the mean of the all of the collected data combined.

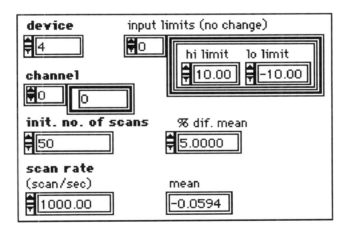

Fig. 11.23 Front panel for MeanComp

The top of the diagram (see Figure 11.24) shows a VI called MeanComp1 connected to all its inputs. MeanComp1 collects the first two groups of data and calculates the mean and the percentage difference of the means. If the percentage difference is less than the specified percentage difference, then the true case of the Case Structure is executed. The variable named mean is the mean of the two groups combined, and it is displayed on the front panel. The front panel and the block diagram for MeanComp1 are described below.

If the percentage difference of the means collected by MeanComp1 is larger than the specified percentage, the false case of the Case Structure is executed. The false case shows all of the connections to MeanComp2, which takes the mean of the two sample records acquired by MeanComp1 as an input. MeanComp2 then acquires another sample record and compares the mean to the input value. MeanComp2 continues to collect progressively longer sample records and to compare means until the two mean values are within the specified percentage. The final mean of all of the sample records combined is displayed on the front panel.

124 Analysis of Sampled Data

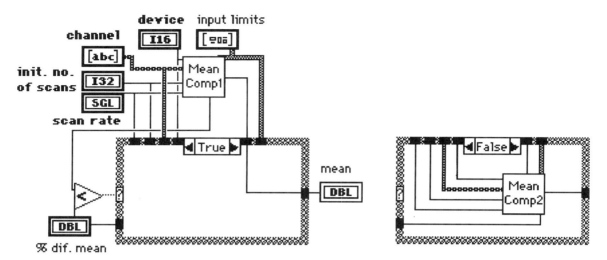

Fig. 11.24 Main diagram for MeanComp

The two subVIs each have their own front panel and diagram. (See Figure 11.25). The front panel for MeanComp1 is similar to the panel for MeanComp with the addition of indicators for the mean value of the combined sample records and the actual percentage difference between the means of the two sample records. Number of Scans is the number of samples in each sample record. The individual means of the two sample records are not displayed.

The main part of the diagram for MeanComp1 is a For Loop, which executes twice. Each time the loop executes, a set of voltage samples is acquired and the mean value calculated. The output of the For Loop is an array containing the two mean values. Within the loop, the loop iteration terminal, i, is wired to the Iteration Input of AI Waveform Scan so that AI Config is called only once. Outside of the loop, the mean of the two sample records combined and the percentage difference between the two means are calculated and displayed.

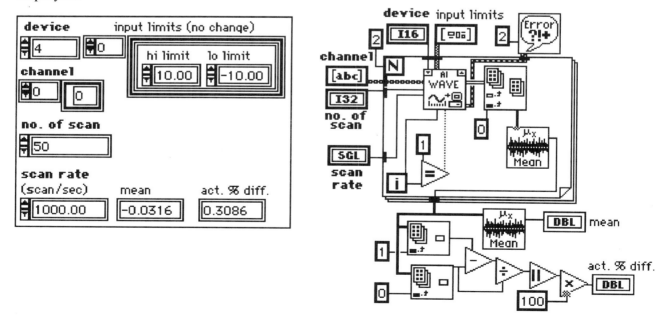

Fig. 11.25 Front panel and diagram for MeanComp1

The front panel for MeanComp2 (see Figure 11.26) is similar to MeanComp with additional inputs to tell it the mean value found by MeanComp1 (init. mean) and the total number of samples in each of the two sample records used by MeanComp1 (init. no of scans). MeanComp2 acquires a new sample record with twice the number of samples as Initial Number of Scans. It then computes the mean value and calculates the percentage difference between this new mean and the initial mean. If the percentage difference is smaller than the specified tolerance, MeanComp2 calculates the average of the two mean values and displays it as the final mean value. If the percentage difference is larger than the tolerance, the sample size is doubled and the process repeated. MeanComp2 is not very useful as a stand-alone VI. It should always be executed in sequence with MeanComp1 as shown in MeanComp.

The main structure of the block diagram is a While Loop. A set of voltage samples is acquired for each iteration with the sample size doubling the previous sample record. The difference of the means and a mean of the entire sample are calculated in the formula node with the mean from MeanComp1 used as the initial value of x1. The new mean is assigned to variable x1, replacing the initial value. The While Loop continues to execute until the difference B is less than the specified percentage. The occurrence of an error also can also stop the loop.

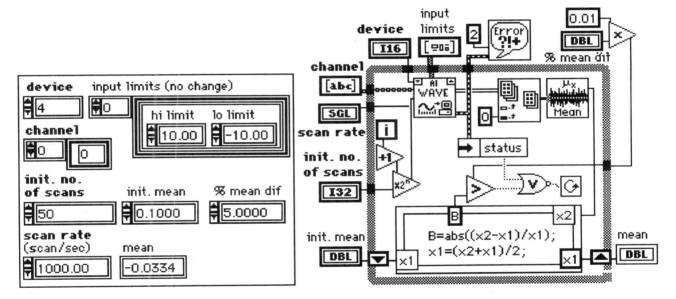

Fig. 11.26 Front panel and diagram for MeanComp2

Confidence Interval for the Variance

The concept of a confidence interval is also very useful when estimating the variance of a random variable. Again, we would like to know, given the size of the sample and the estimated variance, in what range the true variance might be with a given confidence level. We express the confidence interval for the variance as:

$$b_1 s^2 < \sigma_x^2 < b_2 s^2 \qquad a\% \text{ confidence level}$$

where b_1 and b_2 are constants that are less than and greater than unity respectively. Note that we have expressed the confidence interval as a multiplier on the estimated variance and that the confidence interval is not necessarily symmetrical about the estimate.

To evaluate the constants, we must know the distribution function for the variance estimator. It is shown in statistics textbooks that the relevant distribution function is the chi-squared distribution with n degrees of freedom where n = N-1 and N is the number of samples in an individual sample record.

The chi-squared variable with n degrees of freedom is the sum of the squares of n independent random variables, each of which is normally distributed with a mean value of zero and unit variance.

$$\chi_n^2 = z_1^2 + z_2^2 + \ldots\ldots + z_n^2$$

The chi-squared variable is itself a random variable that has a mean value of n and a variance of 2n. The probability density function is described by a complicated function, which is tabulated for various values of n in tables of mathematical functions. What we actually need are the values of χ^2_n for which the integral under the probability density function is equal to a given probability. We then define $\chi^2_{n:a}$ such that:

$$\int_{\chi^2_{n:a}}^{\infty} p(\chi_n^2)\, d\chi_n^2 = a.$$

For example, the probability that χ^2_n is greater than $\chi^2_{n:.1}$ is 0.1. The values of $\chi^2_{n:a}$ for various values of n and a are compiled in statistics textbooks for fairly small values of n. For larger values of n, an approximate formula given below may be used.

The values of $\chi^2_{n:a}$ may be approximated by the formula below for values of n greater than about 30.*

$$\chi_{n:\beta}^2 \cong n\left[1 - \frac{2}{9n} + z_\beta\sqrt{\frac{2}{9n}}\right]^3$$

Here z_β are values obtained from normal distribution tables. A brief tabulation is given in Figure 11.27.

ß	z_β	ß	z_β
0.005	2.58	0.995	-2.58
0.025	1.96	0.975	-1.96
0.050	1.65	0.950	-1.65
0.250	0.675	0.750	-0.675

Fig. 11.27 Table of normal distribution for selected values

We can finally write the confidence interval for the variance as:

$$\frac{ns^2}{\chi^2_{n:a/2}} \leq \sigma_x^2 \leq \frac{ns^2}{\chi^2_{n:1-a/2}} \quad \text{(1-a) confidence level.}$$

---

* From Abramowitz and Stegun, *Handbook of Mathematical Functions* (Dover, 1964).

For example, assume that a set of 1,000 independent samples of a random variable was acquired and the sample variance was 3.45. We wish to find the confidence interval for a 95 percent confidence level, so we select a=0.05. We first compute:

$$\chi^2_{999:0.025} = 999\left[1 - \frac{2}{9(999)} + 1.96\sqrt{\frac{2}{9(999)}}\right]^3 = 999(1.090)$$

$$\chi^2_{999:0.975} = 999\left[1 - \frac{2}{9(999)} \pm 1.96\sqrt{\frac{2}{9(999)}}\right]^3 = 999(0.941)$$

We are now ready to plug into the main formula:

$$\frac{999(3.45)}{999(1.090)} \le \sigma_x^2 \le \frac{999(3.45)}{999(0.914)}$$

or simplifying:

$$3.17 \le \sigma_x^2 \le 3.77 \qquad 95\% \text{ confidence level.}$$

It is important to notice that the confidence interval is not symmetrically distributed around the sample variance. That is, the lower bound of the confidence interval is 0.28 less than the sample variance, while the upper bound is 0.32 greater than the sample variance.

The Uncer VI

Figure 11.28 shows the front panel for a virtual instrument called Uncer, which calculates the mean and the standard deviation confidence intervals. The front panel controls are the same as Vave with the addition of specification of the confidence level. The mean and the standard deviation and their uncertainties are displayed. Parameters used to calculate the uncertainties are also displayed.

The top part of the diagram in Figure 11.29 is Vave connected to all its inputs from the front panel. The outputs, mean and standard deviation, are displayed on the front panel

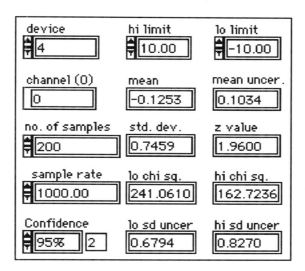

Fig. 11.28 Front panel for Uncer VI

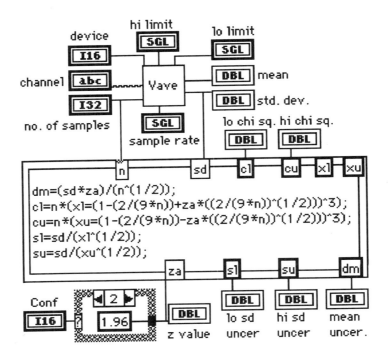

Fig. 11.29 Block diagram for Uncer VI

and also used to calculate the uncertainties. The formula node contains the equations we have just developed. The case structure selects the appropriate z value, which is determined by the confidence level. The confidence parameters are displayed on the front panel.

11.5 Higher Moments of the PDF

The mean value and variance (standard deviation squared) may both be thought of as moments of the probability density function. For example, the mean may be written as:

$$\mu_x = \int_{-\infty}^{\infty} x p(x) dx.$$

and the variance as:

$$\sigma_x^2 = \int_{-\infty}^{\infty} [x - \mu_x]^2 p(x) dx.$$

In general, the nth moment of a probability distribution is defined as:

$$M_n = \int_{-\infty}^{\infty} [x - \mu_x]^n p(x) dx.$$

The higher moments are moments calculated with n greater than 2. Any level of moment can be calculated from a given data record, but the most frequently used higher moments are the third and fourth moments, described below.

The expression given above for the higher moments may be difficult to implement for a typical sample data record where the PDF may not be well resolved. The higher moments are usually calculated from the data in the same way that the mean and the standard deviation are calculated. For example, the third moment is calculated as:

$$M_3 = \frac{1}{N} \sum_{i=1}^{N} (x_i - \bar{x})^3.$$

The third moment can also be calculated from accumulated sums as we did previously for the variance:

$$M_3 = \frac{1}{N} \left[\sum_{i=1}^{N} x_i^3 - 3\bar{x} \sum_{i=1}^{N} x_i^2 + 2N\bar{x}^3 \right].$$

The odd moments reflect the asymmetry of the distribution. The third moment is used to indicate if the distribution is skewed towards positive or negative values. It is zero if the distribution is symmetric and takes on large values if the distribution is highly skewed to one side or the other, as illustrated below. For example, a distribution with a long tail on the positive side of the mean has a large positive third moment. The third moment is usually normalized by the cube of the standard deviation and is called the skewness. (See Figure 11.30.)

$$Sk = M_3/\sigma_x^3$$

The normalized fourth moment is called either the flatness factor or the kurtosis.

$$k = M_4/\sigma_x^4$$

The flatness factor gives some information about the shape of the PDF. The flatness factor is 3 for a Gaussian distribution. Larger values indicate a relatively flat distribution, while small values indicate a peaky distribution with broad tails. The sketches in Figure 11.31 show three distributions which have approximately equal standard deviations but different values of the flatness factor.

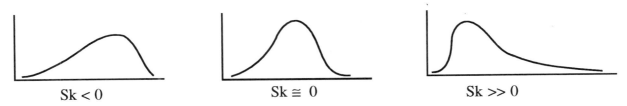

Fig. 11.30 Three PDFs showing negative, zero, and positive skewness

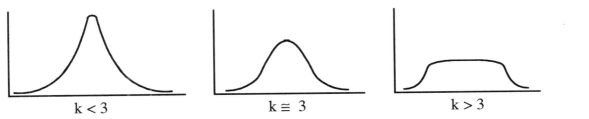

Fig. 11.31 Three PDFs showing various flatness factors

11.6 Correlation

Sometimes we are interested in the relation between two random variables. Such a relationship may be causal, meaning that two variables are related to each other because changes in one cause changes in the other. For example, if the two random variables were the intensity of sunlight striking a detector and the local air temperature, we might expect that the variables would be related. Variations in the intensity of the sunlight cause (at least partially) the variations in temperature. If we were to plot the two variables as a function of time, we would see that the two curves would appear similar. However, the air temperature is also affected by many other factors that are not related to the intensity of sunlight, and we could not establish a direct relationship between the two variables. We need a statistical tool to tell us how strongly the strength of the incident sunlight controls the air temperature.

In other cases, two variables may be related because they are both being affected by the same mechanism. We again select an example from the natural environment. Consider two gauges measuring the height of the water surface, one mounted in an estuary and one mounted on a platform in the ocean several miles offshore from the estuary. The two variables are related because the water surface level at both locations moves up and down with the tide. However, the ocean surface level also varies as waves pass by the measurement location. The surface level in the estuary changes as the river flow changes. Again, we need a statistical tool to tell what fraction of the variation at one location is related to the variation at the other location.

The appropriate statistical tool is a correlation or correlation coefficient. This is a quantitative tool to measure the degree to which the variations of two signals are related. We will find via examples that the correlation has widespread application in studies of both the natural and the technological worlds. It may be used to detect relationships between two random variables, to measure the physical dimensions of spatially varying phenomena, and to determine the duration of temporally varying phenomena.

In the paragraphs below, we first define the cross correlation, or more precisely, the cross covariance. We then show by example how it may be used to study natural phenomena.

The cross-correlation of two random variables $x(t)$ and $y(t)$ is defined as:

$$R_{xy} = \lim_{T \to \infty} \frac{1}{T} \int_0^T x(t) y(t) dt.$$

Usually we are more interested in studying how the fluctuations about the mean of the two signals are related. In that case, we are interested in the cross-covariance, which is defined as:

$$R_{xy} = \lim_{T \to \infty} \frac{1}{T} \int_0^T [x(t) - \mu_x][y(t) - \mu_y] dt.$$

Note that the same symbol is used for both the cross-correlation and the cross-covariance. When most people refer to cross-correlation, they really mean the cross-covariance. It is not immediately obvious how this expression can tell us whether two signals are related. We will find that large positive values of Rxy indicate that the signals tend to move together. Large negative values indicate the signals are still related, yet vary in opposite directions. Small values of Rxy indicate that the two signals are unrelated.

Our first example is for two signals that are fairly strongly correlated The two random variables $x(t)$ and $y(t)$ are shown in Figure 11.32. It is clear that they are not exactly the

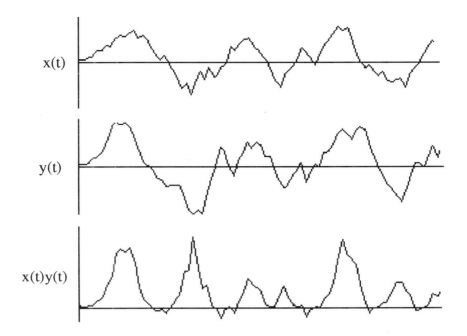

Fig. 11.32 Example of two signals that are strongly correlated

same but are fairly closely related. The major swings in the two signals usually coincide. The product of the two signals is almost always positive, with only small negative excursions. This is true because when x is positive, y is usually also positive, and vice versa. The integral of the product over the duration of the sample record would obviously be a large positive number, indicating that x and y are strongly correlated.

The next example is two signals that are "anti-correlated." (See Figure 11.32.) This means that the two signals are still strongly related, but when x varies toward the positive, y varies toward the negative. Therefore, the product of the two signals is nearly always negative, and the integral of the product over the entire sample record is a large negative

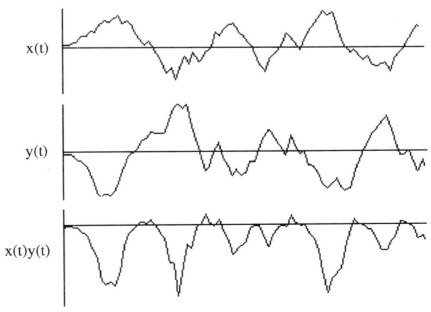

Fig. 11.33 Example of two signals that are anti-correlated

132 Analysis of Sampled Data

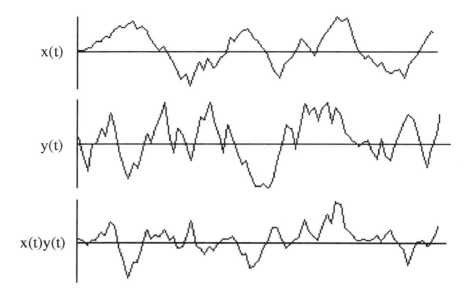

Fig. 11.34 Example of two signals that are uncorrelated

number. We say that these signals are negatively correlated or we might also say they are moving in anti-phase.

The final example shows two signals that are uncorrelated, meaning that there is no relation between them. (See Figure 11.34.) We see that a peak in y(t) is just as likely to correspond to a peak as to a valley in x(t). The product varies between positive and negative values, and it appears that the integral is a small value. Thus, we see that the cross covariance as defined above is a good indicator of the degree of correlation between two signals. So far, though, we have not spoken quantitatively about the degree of correlation.

The cross-covariance is usually normalized by the product of the standard deviations of the two signals. The normalized cross-covariance is then called the correlation coefficient.

$$\rho(x,y) = \frac{R_{xy}}{\sigma_x \sigma_y}.$$

The correlation coefficient is very useful for determination of how closely two signals are related. The maximum value of the correlation coefficient attained if the two signals are exactly correlated is 1. A correlation coefficient larger than about 0.5 means that the two signals are strongly correlated. The most negative possible value is -1, and negative correlation coefficients around -1 indicate that the signals are strongly anti-correlated. Small values of the correlation coefficient (less than about 0) indicate that the two signals are either very weakly correlated or completely uncorrelated. Thus, by normalizing the cross-covariance, we can now say a considerable amount about the relationship between two signals, regardless of the amplitude of the individual signals.

Measurement of the Cross-Covariance

With computerized data sampling, continuous data records of the two signals are not available. What we do have available are sampled data records. The cross-covariance is calculated from a pair of sampled data records as:

$$R_{xy} = \frac{1}{N} \sum_{i=1}^{N} (x_i - \bar{x})(y_i - \bar{y}).$$

where $\bar{x}$ and $\bar{y}$ are the sample means for the x and y variables.

The requirements on our sampling are somewhat more stringent than they were for single-point statistics. In particular, the definition requires that pairs of samples be acquired simultaneously. This cannot be done with an ordinary multifunction board like the National Instruments NB-MIO-16, which has only a single sample-and-hold and a single A-to-D converter. Simultaneous samples can be acquired with high-performance data acquisition cards from several manufacturers. For example, the National Instruments NB-A2000 has four separate analog input channels each with a sample-and-hold. The sample-and-holds can be triggered together to achieve simultaneous sampling.

The statistical uncertainty in the cross-covariance estimate goes down as the reciprocal of the square root of the number of samples. You can understand this by comparing the formula above to the estimator of the variance given previously. The individual sample pairs must be independent from other sample pairs in order to minimize the uncertainty. Just as in measuring the variance, you want to sample as slowly as is practical. If the measurements of the cross-covariance are not repeatable to an acceptable level, you should first try decreasing the sampling rate. If a decrease in sample rate does not improve the repeatability, you should then increase the total number of samples in the record.

The above discussion assumes that you have the appropriate hardware in your computer to sample two signals simultaneously. Often such hardware is not available. In such situations, the best you can do is sample the two signals alternately. In this case, it is necessary to sample as rapidly as possible. This is illustrated as we develop our virtual instruments for measuring the correlation coefficient.

The Corr VI

We will now develop a VI called CORR to calculate the correlation coefficient between two voltage signals sampled by two channels of the A-to-D converter. We will use the standard hardware described in this tutorial, an NB-MIO-16 multifunction board. There is only a single sample-and-hold on this board, so we cannot obtain simultaneous samples from the two channels. We must then sample the signals as rapidly as possible to have the minimum time lag between sampling the two channels. The NB-MIO-16 samples its ADC at a uniform rate. That is, if the sample rate is 50,000 per second for each of two channels, the time interval between nominally simultaneous samples of the two channels will be 1/100,000th of a second. We must sample rapidly enough that the time lag is insignificant. The problem with sampling rapidly is that we get a lot of data in a very short time.

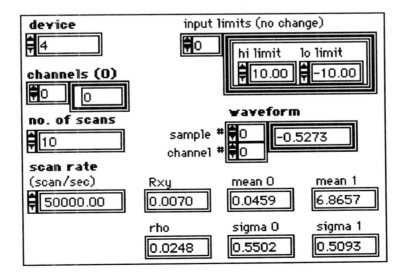

Fig. 11.35 Front panel for the Corr VI

134 Analysis of Sampled Data

Normally, we would like to acquire samples over a relatively long time in order to get statistically converged results. However, the calculation time increases as the square of the number of samples. In using the program CORR it will thus be important to repeat the correlation measurement several times and to average the resulting correlation coefficients.

The front panel is similar to AI Waveform Scan described in section 8. Outputs of the raw data samples, the mean and the standard deviation of each signal, the cross-covariance, and the correlation coefficient are displayed. (See Figure 11.35.)

There are four main steps in the diagram for the Corr VI. First, in frame 0, the array of voltage samples is acquired using the AI Waveform Scan VI, and the mean and the standard deviation are computed for each signal. The summation needed to calculate the cross-covariance is done in the For Loop in the top half of frame 1. The variable z holds the sum. The cross-correlation and the cross-covariance are calculated in the Formula Node in the bottom of the frame, using the result from the For Loop. This node divides the z value by the number of scans and normalizes the result by the product of the standard deviations. (See Figure 11.36.)

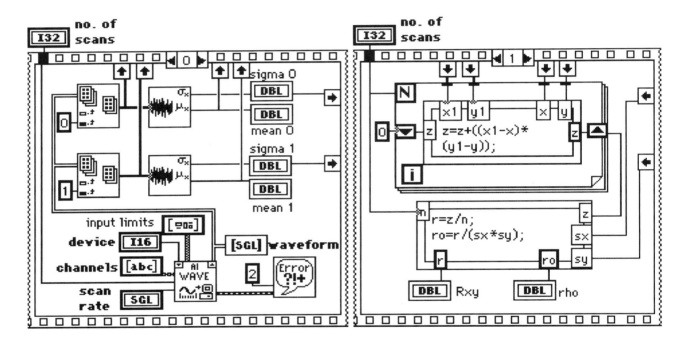

Fig. 11.36 Diagram for the Corr VI

An Alternative VI for Measuring Correlations

We now develop a new VI called Corr2, which makes better use of the available hardware to measure the correlation. The problem with Corr is that in order to sample the two channels nearly simultaneously, we had to use an overall sampling rate that was too high. Normally, we would like to acquire samples over a relatively long time to get statistically converged results. Thus, use of Corr would require an inordinately large number of samples.

The Corr2 VI uses AI Read One Scan VI within a For Loop to sample repeatedly the two voltage signals. Inside the loop, AI Read One Scan samples the two channels at an interval of about 37 microseconds* using an NB-MIO-16L. The outer loop then controls the overall sampling rate.

This VI is excellent if you are measuring correlations between relatively slowly varying signals, say, maximum frequencies less than about 5,000 Hz. If you have rapidly varying signals, you will have no alternative but to purchase an acquisition board that allows true simultaneous sampling.

The front panel for Corr2 (see Figure 11.37) is the same as for Corr except that there is no control for the scan rate. The Wait Period controls the overall sampling rate, which is the rate that the For Loop is executed. Wait Period is specified in multiples of milliseconds. The minimum value should be 50, which means that the For Loop will cycle every 50 milliseconds. The Output Array contains the actual data samples. Rxy and Rho are the covariance and the correlation coefficient respectively. Finally, the mean and the standard deviation are displayed for each of the two channels.

The diagram is similar to Corr except that the data are collected using AI Read One Scan. The bottom of frame 0 shows the AI Read One Scan VI connected to its inputs inside a For Loop. The Wait Till Next ms Multiple function controls the overall sampling rate. The mean and standard deviation values are calculated for each channel at the top of the frame. The values are sent to frame 1. Frame 1 of Corr2 is identical to frame 1 of Corr, so it is not shown here.

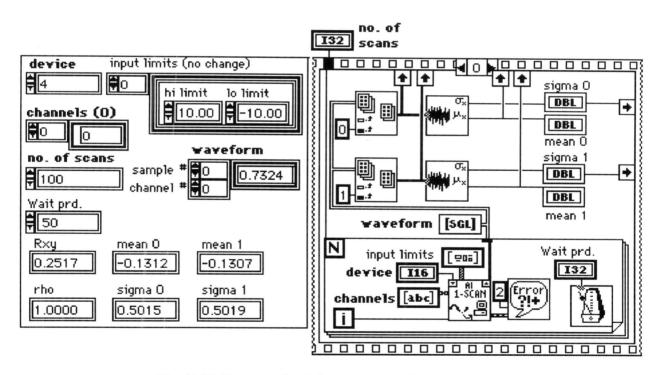

Fig. 11.37 Front panel and diagram for the Corr2 VI

---

* Note that the interchannel delay may be dependent on the exact board and computer model you use. We measured it by reading the same triangle wave with the two ADC channels as described in Chapter 8.

Examples of Application of the Cross-Correlation

Our first example of the use of the cross-correlation is an experiment with human subjects. We may be interested in factors that influence a person's skin temperature. Suppose that we have made the hypothesis that there is a direct connection between noise and skin temperature. The idea is that noise causes stress and raises the skin temperature. To investigate, this we could attach a skin temperature sensor to a subject and expose him to different types of noises. The temperature sensor would supply one voltage signal, while the other signal would come from a sound pressure level gauge.

We would probably want to set up Corr2 to run with a very low sampling rate since the experiment would run over a long period of time. We would then expose the subject to various noise levels while simultaneously providing other stimuli that might also affect the skin temperature. For example, we might vary the room temperature, the number of people in the room, or the difficulty of the task the subject is doing. If noise level really does have a significant effect on skin temperature, then we would expect a correlation coefficient well above 0.

For our second example, we consider a rotating device driven by a DC motor via a belt drive. An engineer studying speed variations of the device notes that the torque delivered to the driven shaft fluctuates. This could be caused by various problems, including failing bearings, loose belts, and fluctuating external torques. The engineer suspects that the voltage supplied by the motor controller is not steady and is causing the fluctuations. To test this, the A-to-D is set up to sample the motor voltage with one channel and the torque via a torque sensor on the driven shaft with another channel. When the measurement is complete, the correlation coefficient is 0.1. This suggests that the voltage fluctuations are responsible for only a small fraction of the torque fluctuations. The engineer will have to look into other sources to find the primary cause of the problem.

The final example involves using the correlation coefficient to measure the physical scale of wind gust events. The question can be phrased: over how great a distance do velocity fluctuations remain correlated? To measure this we would use two wind velocity sensors capable of giving a signal proportional to the instantaneous velocity at the sensor position. We would then measure the correlation coefficient for different separation distances between the sensors. A typical plot of correlation coefficient vs. separation distance is shown in Figure 11.38. At small separations the correlation coefficient is near 1. This shows that the instantaneous velocity is almost the same at closely spaced points. The correlation coefficient falls off rapidly. This rolloff corresponds to smaller-scale eddies in the atmospheric turbulence. The correlation drops to zero at large separations, indicating that only an infrequent gust is as large as 20 meters, with most in the 5-to-20 meter range.

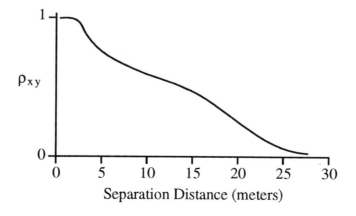

Fig. 11.38 The correlation coefficient between two different wind sensors

Cross-Correlation Function

It is often the case that we wish to correlate two signals with a time delay applied to one of the signals. In that case, we usually use the cross-correlation function defined as:

$$R_{xy}(\tau) = \lim_{T \to \infty} \frac{1}{T} \int_0^T [x(t) - \mu_x][y(t+\tau) - \mu_y]dt.$$

Here we see that the cross-correlation is not just a single value but is a function of a time delay, τ, applied to one of the signals. By inspection we can see that the maximum value of the cross-correlation function is still the product of the standard deviations of the two signals. Therefore, the cross-correlation function is usually normalized by the standard deviations to form a cross-correlation coefficient that is a function of the time delay, τ.

$$\rho_{xy}(\tau) = \frac{R_{xy}(\tau)}{\sigma_x \sigma_y}.$$

A plot of the cross-correlation coefficient as a function of time delay is called a cross-correlogram. A typical cross-correlogram is sketched in Figure 11.39. Notice that the function may range between -1 and 1 and that the peak value does not necessarily occur at 0 time delay.

The cross-correlogram has many uses, including detecting correlation when none exists at zero time delay, measuring time delays, and extracting characteristic patterns from noisy signals. Some of these applications are illustrated below.

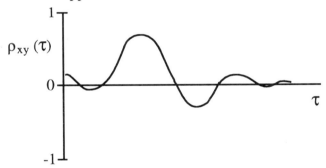

Fig. 11.39 A cross-correlogram

Measurement of the Cross Correlation Function

We assume that we have two data arrays, called x and y, that were acquired simultaneously. That is, x_0 and y_0 were acquired at exactly the same time. The time interval between successive samples in a data array is Δt and there are n samples in each array. Then the cross correlation function can be computed only at specific time delays, namely:

$$\tau_k = k \, \Delta t \qquad \text{where } -n < k < n.$$

The cross-correlation is computed by forming the sum of the products with one array shifted by k positions. Note that there are only n-k sample pairs available when one array is shifted by k positions. The full formula is then:

$$R_{xy}(k \, \Delta t) = \frac{1}{n-k} \sum_{i=1}^{n-k} (x_i - \bar{x})(y_i - \bar{y}) \qquad -n < k < n.$$

Normally the cross-correlation is computed at all available time delays (a total of 2n-1 different values). You can see that the computation of the cross correlation function will be time-consuming if the data arrays are large.

A simple example is shown below to help you understand the formula. Here the two data arrays each contain six elements. The chart on the left illustrates the computation of the cross-correlation at zero time delay. The value from the y array is multiplied by the corresponding value from the x array. The six products are summed; then the sum is divided by the total number of available pairs (6) to form the correlation. The chart on the right illustrates the computation with a time delay of two sample intervals. The y array is shifted by two positions and the products formed again. In this case there are only four sample pairs in the sum. Dividing the sum by 4, we see that the correlation is considerably stronger with the time delay applied.

Cross-correlation with zero time delay

i	x_i	y_i	$x_i y_i$
1	2	3	6
2	3	1	3
3	4	2	8
4	3	3	9
5	2	4	8
6	1	3	3

$\sum x_i y_i = 37$

$R_{xy}(0) = 37/6 = 6.17$

Cross-correlation with time delay of 2 samples intervals

i	x_i	y_i	$x_i y_i$
1	2	2	4
2	3	3	9
3	4	4	16
4	3	3	9
5	2		
6	1		

$\sum x_i y_{i+2} = 38$

$R_{xy}(2\Delta t) = 38/4 = 9.5$

There are practical considerations in measuring the cross-correlation function. The most important issue is that we now have to sample fast enough to achieve the shortest time delay of interest. For example, if we are interested in the cross correlation function for time delays ranging from 0.001 seconds up to 0.1 second, we have to sample each input at 1,000 Hz. We also have to acquire enough samples to obtain a reasonable estimate of the correlation. In the present example, we need at least 200 samples of each channel.

Since we are now acquiring data rapidly, our sample period may be too short to obtain statistically converged results. Again sticking with the example, our total sample period would be 200 samples X 0.001 sec./sample = 0.2 seconds. We could acquire a very long data record, say, 10,000 samples, but the computation time for the cross-correlation function would be prohibitive. The appropriate thing to do would be to acquire several records of 200 samples. The cross-correlation functions calculated from the multiple records would then be averaged.

CorrFun VI

The front panel for CorrFun VI, which calculates the time-delayed cross-correlation coefficient and plots the result, is shown in Figure 11.40. The front panel for CorrFun is similar to that for Corr. The cross-covariance and the correlation coefficient are a function of time and displayed as an array. The actual VI displays the graph of the correlation coefficient as a function of time delay on the panel. The graph is omitted here to save space.

The frame 0 is very similar to Corr. The AI Waveform Scan VI acquires the voltage samples, and the mean and the standard deviation are calculated for each channel. The product of the standard deviations is calculated and sent to frame 2.

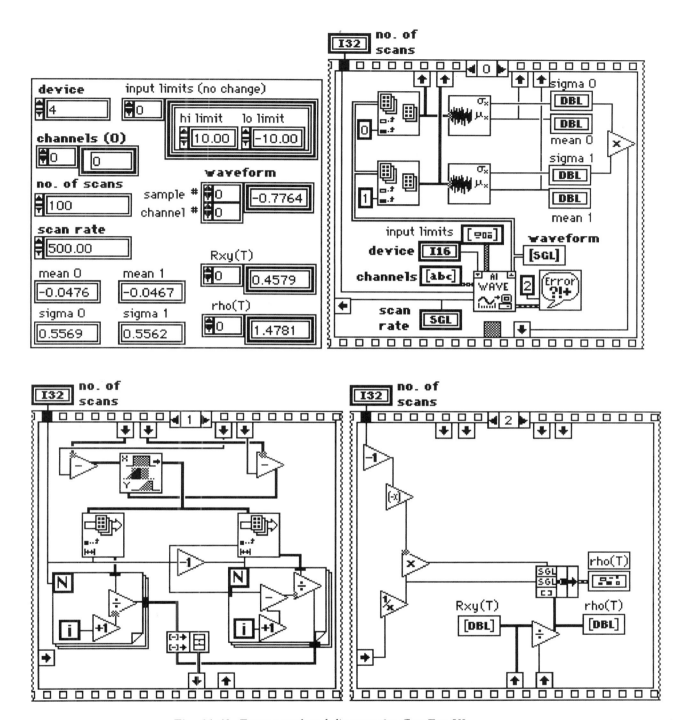

Fig. 11.40 Front panel and diagram for CorrFun VI

In frame 1, the mean is subtracted from each data sample and then the data arrays are sent to the Cross Correlation function which does the multiplication and summation. This function calculates the correlation for all time delays ranging from $-(n-1)\Delta t$ to $(n-1)\Delta t$. The LabView Cross Correlation function does not divide the result by the number of sample pairs in the sum. The bottom of frame 1 does this for you. The number of available sample pairs is different for each time delay. The first element of the cross correlation array corresponds to a time delay of -(n-1)Δt. There is only one available sample pair for this time shift. There is one additional sample pair available for each successive element of the correlation array until zero time delay, when there are n sample pairs. Continuing through the cross correlation array, the number of sample pairs decreases by one for each increment in the time delay. The left For Loop in frame 1 does the division for the first half of the correlation array, while the right loop does the second half of the array. The two arrays containing the cross-correlation coefficients are put back into one array, using the Build Array VI, and sent to frame 2.

The cross-correlation coefficient values are normalized by dividing by the product of the standard deviations of the voltage samples from the two channels. The result is displayed in the array labeled rho(T) and as a graph on the front panel. To determine the correct scale for the x-axis of the graph, the actual sampling time interval must be determined. This is calculated as the reciprocal of the scan rate.

Cross-Correlation Function Examples

As an example of the time-delayed correlation we consider a unique experiment to measure the speed of sound. The experimental apparatus consists of two microphones, a personal computer, and a person lighting firecrackers. The computer version of this book shows an animation of a signal that results each time a firecracker explodes. (See Figure 11.41.)

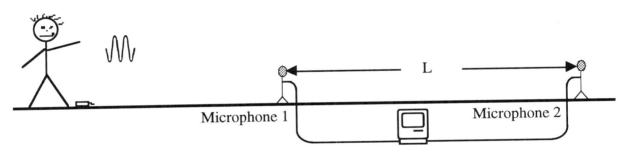

Fig. 11.41 Cross-correlation function example 1

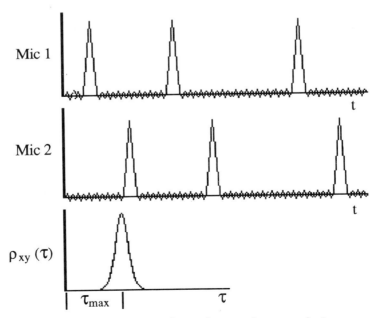

Fig. 11.42 Signals and correlogram for example 1

The signal at each microphone consists of random noise, with an occasional large pulse from a firecracker exploding, as shown in Figure 11.42. The pulses are randomly spaced in time, but the time delay between the pulses arriving at the two microphones is always the same. The correlogram shows a strong peak at the time delay corresponding to the transmission time of the sound wave from microphone 1 to microphone 2. In order to calculate the speed of sound, one simply divides the distance between the two microphones by the time delay for maximum correlation.

Extracting Signal from Noise

In this example we will demonstrate how a specific pattern may be found in a signal that is contaminated by noise. Assume that we are transmitting a pulse across a very noisy transmission path. We know the pulse shape, but we also want to know exactly when it was received. Normally, we would detect a pulse with a level detector, but because of the severe noise contamination we would get many false detections. The sketch in Figure 11.43 shows a segment of the signal as received at our computer. The small diagram at the top of the figure shows the pulse as transmitted. We can find the same pulse hidden in the noisy record (see the arrow) but how can we recognize it with a computer program?

The first step in extracting the pulse is to digitize the incoming signal with the A-to-D converter. The digitized record is shown in Figure 11.44. We also digitize the known shape of the transmitted pulse. In this case, we have used 12 samples to represent the pulse. Note that the time interval between samples must be the same for both the sampled data record and the pattern function. The cross-correlation between the sample data record and the digitized pulse shape is then calculated. This may be thought of as sweeping the pulse shape along the sample record as illustrated by the animation in the computer version of *LabTutor*. The animation pauses at the correct time delay, and you can see that the signals line up closely. The cross correlation is generally small except at the time delay, where the sampled data and the transmitted data pulse shape align. Therefore, we could use this technique to determine when the pulse arrived.

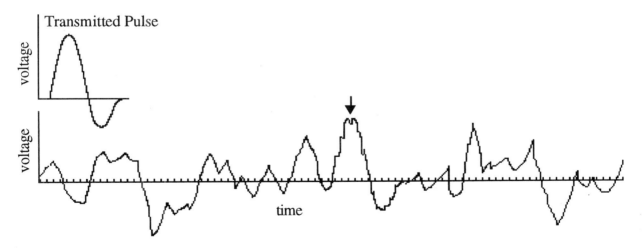

Fig. 11.43 The transmitted pulse on the top and the same pulse shown buried in a noisy signal

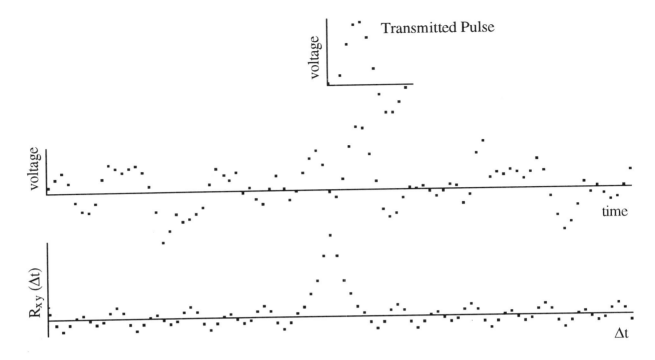

Fig. 11.44 Digitized signal and computed cross correlation

We now examine a situation in which the correlation function is used to isolate the source of a vibration. The situation is that an external vibration is interfering with the operation of a precision machine. There are several pieces of heavy equipment nearby that may cause the vibration. To check if a given machine is the source, an accelerometer is mounted on the floor adjacent to the precision machine and another one on the potential troublemaker. If we have found the source, then we expect a strong correlation between the signals from the two accelerometers. However, since the vibrations at the two measurement points are not necessarily in phase, we can find the correlation only by measuring the full cross-correlation function. In this case we expect the correlogram to look like a decaying sine wave as shown in Figure 11.45. Note that if we had only looked at the correlation at zero time delay we might not have discovered the problem.

Autocorrelation Function

The term *autocorrelation* means a correlation of a signal with itself. This is a logical extension of the cross-correlation with time delay. The idea is to determine how well a signal is correlated with itself at a later time. The utility of this will become evident after a few examples. Formally, the definition of the autocorrelation is:

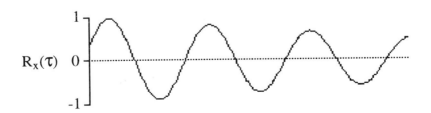

Fig. 11.45 Correlogram measured in the vibration example

$$R_x(\tau) = \lim_{T \to \infty} \frac{1}{T} \int_0^T x(t)\, x(t+\tau)\, dt.$$

As usual, we are more interested in the autocovariance:

$$R_x(\tau) = \lim_{T \to \infty} \frac{1}{T} \int_0^T [x(t) - \mu_x]\, [x(t+\tau) - \mu_x]\, dt.$$

Usually when people talk about the autocorrelation they are really referring to the autocovariance. As with the cross-correlation, it is convenient to talk about an autocorrelation coefficient, defined as:

$$\rho_x(\tau) = \frac{R_x(\tau)}{\sigma_x^2}.$$

where R_x is the autocovariance.

A typical plot of the autocovariance coefficient vs. time delay is shown in Figure 11.46. There are several features to notice. First, the value of the autocovariance coefficient at 0 time delay is just 1. This just says that a signal is exactly correlated with itself. Second, the correlation can have both positive and negative values. It is easy to see why the correlation can be negative by imagining the autocorrelation of a sine wave. Finally, you should note that the autocovariance coefficient is plotted only for positive time delays. This is so because the autocorrelation is an even function, meaning that the value of function is the same for either positive or negative time delays. You should reexamine the definition of the autocorrelation to convince yourself of this. Of course, we could compute and plot the entire autocorrelation function, but it would be a waste of time.

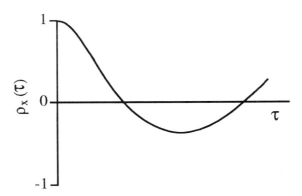

Fig. 11.46 A typical plot of the autocorrelation coefficient vs. time

To examine the utility of the autocorrelation function, we consider a random signal that has a range of time scales over which variations occur. For example, consider once again the fluctuations of the wind velocity on a gusty day. We expect that two velocity samples separated in time by one millisecond will be the same. Stated in another way, the autocorrelation with a time delay of one millisecond is one. On the other hand, if we were to measure the autocorrelation with a delay of five minutes, we would expect no correlation. A measurement of the autocovariance function would allow us to be much more precise. The plot in Figure 11.47 is typical for a situation like this. The fact that the curve remains flat up to a delay of about two seconds indicates that the most rapid

144 Analysis of Sampled Data

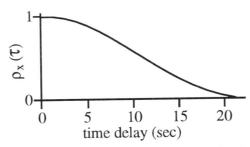

Fig. 11.47 The autocorrelation of the wind velocity

variations occur at a time scale greater than two seconds. The function rolls off rapidly between 5 and 15 seconds, suggesting that most of the fluctuations have time scales in that range. The correlation dies out beyond 20 seconds, indicating that gusts of longer duration are very rare.

The gusty wind example above showed an autocovariance function that decayed monotonically from unity at small time delay to zero at large time delay. This is characteristic of a completely random signal with a range of time scales. A periodic signal, though, would have a dramatically different autocovariance function, as demonstrated for a sine wave in Figure 11.48. The autocorrelation of a periodic function is itself periodic. The signal is exactly correlated with itself whenever the time delay is an integer multiple of the oscillation period.

The autocorrelation of a perfectly periodic signal is not too interesting. Since the signal is fully defined, we have no reason to examine it statistically. However, we should recognize the signs of a periodic signal, since autocorrelations may often show a hint of an imbedded periodic signal in what may appear to be a fully random signal. Sometimes a signal is not truly periodic but still exhibits a strong component that is nearly periodic. The periodic wave may be contaminated by noise, or the period and waveform may vary slightly from cycle to cycle. A classic autocovariance function for such a signal is sketched in Figure 11.49. We see that the autocovariance shows a periodic oscillation, but the oscillation is contained in an envelope that decays toward zero amplitude. This says

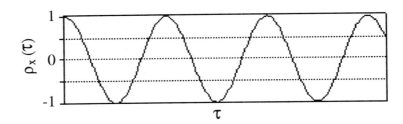

Fig. 11.48 The autocorrelation of a sine wave

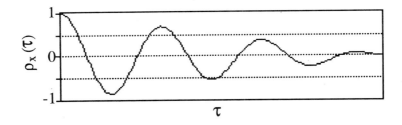

Fig. 11.49 Autocorrelation of a non periodic wave

that the signal is well correlated with itself over a short duration, but over a longer time period the correlation is lost. Small random variations in phase or period from cycle to cycle add up over several periods until, beyond a certain time delay, the phases of the waves are random relative to each other. This means that the signal is no longer correlated with itself. The time it takes the autocovariance to decay to near zero is sometimes called the coherence time.

The situation is often not quite as simple in practice as described. In Figure 11.50, the upper plot shows a short section of a signal that is dominated by a strong oscillation. The waveform and the period vary somewhat from wave to wave, and there is a small amount of random noise. The autocorrelation of this short record is also shown. The autocorrelation oscillates as we expect for a nearly periodic signal. However, in this case the amplitude does not decay monotonically. This is largely the fault of the short record used. In this case, the autocovariance function plot can be used to determine the average period of the wave and to give a rough measure of the coherence time.

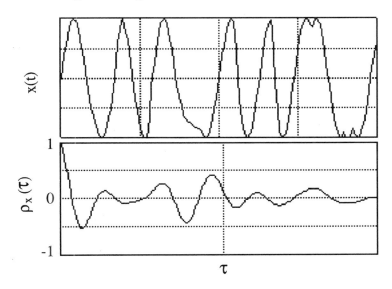

Fig. 11.50 A plot of a periodic wave with noise and it's autocorrelation

Measurement of the Autocovariance Function

The autocovariance function can be computed from an array of data sampled at uniform time intervals by forming the sum:

$$R_x(k \Delta t) = \frac{1}{N-n} \sum_{i=1}^{N-n} (x_i - \bar{x})(x_{i+1} - \bar{x}).$$

This is exactly analogous to the way we computed the cross-correlation. Note, though, that we do not need to compute the autocovariance for negative time delays, since the function is symmetric. LabView includes an autocorrelation VI in the Analysis VIs menu. This VI does not do exactly what you might expect. First, the VI calculates the true autocorrelation, rather than the autocovariance. The easiest way to get the autocovariance is first to subtract the sample mean from each sample of the data array. Also, the VI does not normalize the sum by the number of sample pairs used to form the sum. The number of sample pairs available decreases with increasing time delay . Finally, the VI computes the autocorrelation for both positive and negative time delays. This can make interpretation of the results confusing. The middle element of the results array is the autocorrelation with zero time delay.

When measuring the autocovariance, you must think carefully about the sampling rate and the number of samples acquired. The sampling rate determines the smallest time delay available. The autocovariance will always be unity at zero time delay, and usually you want to be able to resolve the shape of the autocovariance curve. Therefore, it is important to sample fast enough that the autocovariance at the smallest time delay is only slightly less than 1. This can be tested by repeatedly measuring the autocovariance while successively decreasing the sampling rate until the value at the first time delay is unacceptably small.

In considering the number of samples, we should recognize that the number of available sample pairs decreases with increasing time delay. There is only a single sample pair available at the largest possible time delay. Plots of the autocovariance often appear very smooth at short time delays and very noisy at long time delays because of the small sample size available. The best way to get around this is to take a large set of sample records, each just long enough to include the maximum time delay of interest. The autocovariance is then computed for each record and the results averaged.

We will next develop a VI to measure the autocovariance. However, use of this VI is recommended only for relatively short sample records. We will learn below that the autocorrelation function is the Fourier transform of the power spectrum. The power spectrum can be calculated efficiently from sampled data using Fast Fourier Transform (FFT) techniques. It is computationally more efficient to compute the power spectrum of a signal and Fourier-transform it than to compute the autocorrelation directly. The only constraint is that the number of data samples must be an integer power of 2.

Normally, the only time we will calculate the autocorrelation directly is when we have collected the data at uneven time intervals. In this case, the only way to find the power spectrum is to first compute the autocorrelation and then Fourier transform it. This will be discussed in more detail at the end of Section 11.7.

AutoCorr VI

AutoCorr is a VI to calculate and display the autocovariance of sampled data. AutoCorr uses AI Waveform Scan to acquire the data. The front panel has the usual controls for setting up data sampling. The autocovariance is displayed as an array and also in graphical form. The graph displays the autocovariance as a function of time delay. Note that only positive time delays are shown. The actual data, mean, and standard deviation are also displayed. (See Figure 11.51.)

The diagram for AutoCorr is also shown. The bottom of the first frame is the data acquisition part of the VI. It is basically AI Waveform Scan connected to all of its inputs. The output of AI Waveform Scan is a two-dimensional array, so the Index Array function is used to select the column which contains the data samples. The mean is subtracted from each sample before the data array is sent to frame 1. The standard deviation is also calculated and squared before sending it to the next frame. In frame 1 the array of data is processed by the Autocorrelation function. This VI forms the sum for all possible time delays, both negative and positive. Thus, the output has 2n-1 elements, where n is the total number of samples. We take a subset of the output array, since the autocorrelation is symmetric and only positive time delays are of interest. The For Loop is used to normalize the autocorrelation values by the number of samples pairs in the sum. Note that there are n sample pairs for 0 time delay, n-1 pairs for a time delay of one sampling interval, etc. The result is normalized by dividing by the product of the standard deviations and displayed as a graph on the front panel.

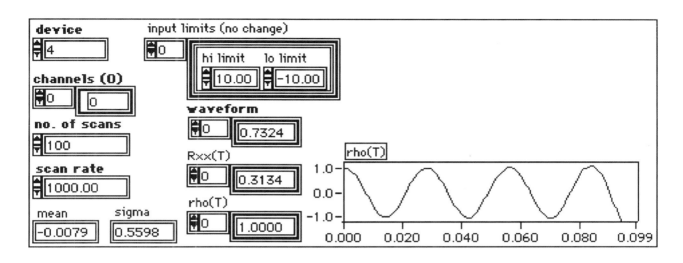
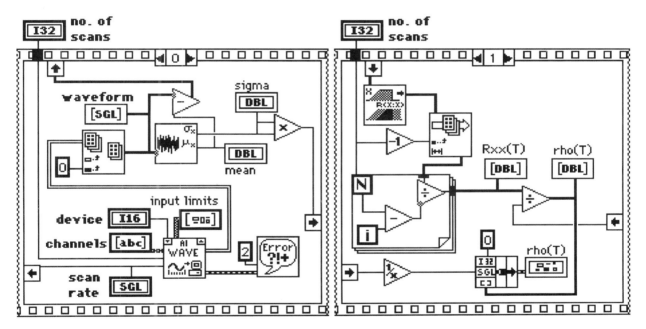

Fig. 11.51 Front panel and diagram for AutoCorr VI

11.7 Power Spectrum

The power spectrum, or, more formally, the power spectral density function, is a way to represent the frequency content of a signal. The power spectrum tells us what fraction of the signal fluctuations occurs in a given frequency band. For example, the power spectrum of a pure tone would tell us that all of the fluctuation power was at a single frequency. For a random signal, the power spectrum would tell us the range of frequencies over which oscillations occurred and the frequency, with the maximum power density.

Formally, the power spectral density function is defined as:

$$G_x(f) = \lim_{\Delta f \to 0} \frac{1}{\Delta f} \left[\lim_{T \to \infty} \int_0^T [x^2(t,f,\Delta f) dt] \right].$$

This expression tells us the power density at a given frequency, f. Breaking down the equation we see that the expression inside the brackets is the mean square value of the signal $x(t,f,\Delta f)$. The function $x(t,f,\Delta f)$ is just the original signal f(t) filtered around the frequency f with a bandwidth of Δf. The integral is then normalized by this bandwidth. To understand this we need to first study filtering.

A filter is a device that eliminates oscillations above or below a specific cutoff frequency. A low-pass filter eliminates any frequency components above the cutoff frequency, while a high-pass filter eliminates frequencies below the cutoff. An example of a low-pass filtering operation is shown Figure 11.52. The signal is a sine wave at 100 Hz contaminated by high-frequency noise. Low-pass filtering with a cutoff frequency of 200 Hz eliminates the noise but leaves the 100 Hz sine wave unaltered. Conversely, if we high-pass-filtered the signal with a cutoff frequency of 200 Hz, the base sine wave would be eliminated, leaving only the high-frequency noise.

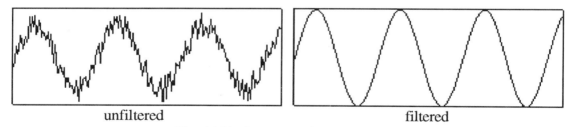

Fig. 11.52 Low-pass filtering operation

A filter is described by a characteristic curve as sketched in Figure 11.53. The gain of the filter is plotted on the vertical axis as a function of frequency. For a low-pass filter, we see that the gain is unity for frequencies well below the cutoff frequency. Around the cutoff frequency, the gain begins to fall rapidly. The cutoff frequency is defined as the frequency at which the power of the signal is reduced by half. We see that the signal is actually attenuated somewhat for frequencies below the cutoff frequency, and some of the signal at frequencies above the cutoff is allowed to pass. The precise shape of the characteristic curve is dependent on the design of the filter. For the present we may consider an ideal filter called a "boxcar" filter which has the ideal characteristic seen in the graph on the right of the figure.

To understand the power spectrum we need to consider something called a bandpass filter. This is a filter that allows signals to pass only within a certain band of frequencies. A bandpass filter may be characterized by lower and upper cutoff frequencies. However, it is more common to characterize the filter by a central frequency of the pass band (f_0 on the plot in Figure 11.54.) and a bandwidth (Δf).

Power Spectrum 149

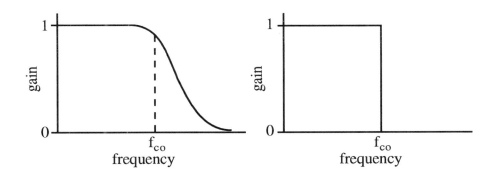

Fig. 11.53 Gain vs. frequency characteristic for a realistic filter (left) and a boxcar filter (right)

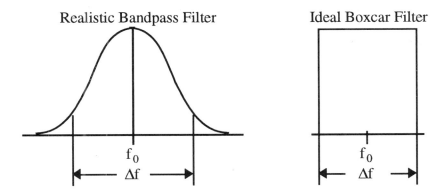

Fig. 11.54 Bandpass filter characteristics

The easiest way to understand the power spectrum is to interpret it in terms of the filtering and measurement operation shown in Figure 11.55. The signal is fed to a bandpass filter with an adjustable center frequency f_0 and a fixed bandwidth Δf. The output from the filter is fed to a device that can measure the mean square voltage. The result is normalized by the filter bandwidth. The setup shown would give the value of the power spectrum at a single frequency f_0. By varying the center frequency of the filter, we could measure the power spectrum for any arbitrary frequency. This is how spectrum analyzers used to work. The center frequency of the bandpass filter was automatically swept through a preselected range of frequencies, and the output from the system was plotted on a chart. The process was very slow, since the system had to dwell long enough at each frequency to measure accurately the mean square value. We will see that modern digital systems are much faster and easier to use.

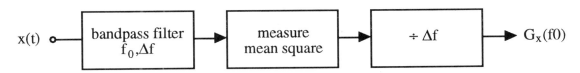

Fig. 11.55 Filtering and measurement operation

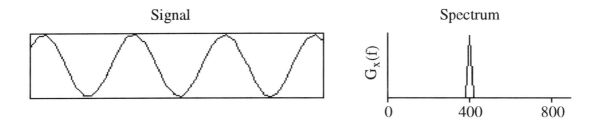

Fig. 11.56 A sine wave and its spectrum

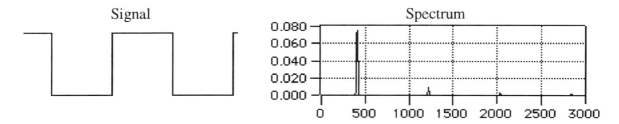

Fig. 11.57 A square wave and its spectrum

At this point it is helpful to examine the spectrum of various signals to develop a better feeling for the meaning and utility of the power spectrum. Each example will show a portion of a time record of a signal and a power spectrum for the signal. In addition, the signal can be played through the computer's speaker in the computer version of *LabTutor*. Our first signal is a pure sine wave. (See Figure 11.56). The power spectrum is just a single spike at the sine wave frequency. The sound you hear is a pure tone at 400 Hz. This example shows you that if you see a spectrum dominated by a large spike, there is a strong pure tone in the signal.

We next study a square wave, which is chosen as representative of non-sinusoidal periodic functions. (See Figure 11.57). A periodic function usually has a strong spectral peak at the fundamental frequency. All the power not contained in this peak is distributed among the harmonics, that is, frequencies that are integer multiples of the fundamental frequency. In the case of the square wave, we see that there is power only in the odd harmonics. The sound you hear is for a square wave at 400 Hz. To get a better feel for the harmonics, we also play the same signal bandpass filtered around 1200 Hz. You should be able to hear the decrease in amplitude for the successive harmonics, as illustrated on the power spectrum.

Any periodic signal may be represented in terms of sine waves at integer multiples of the fundamental frequency. This is called a Fourier series representation of the signal. Mathematically, this is written as:

$$x(t) = \sum_{n=1}^{\infty} a_n \sin(n\omega t + \phi_n).$$

where a_n is the amplitude of the nth harmonic and ϕ_n is the phase angle. A power spectrum is essentially an analysis of a signal in terms of sine waves. Since a periodic signal can be represented as a sum of contributions at discrete frequencies, the power spectrum will show power only at discrete frequencies. Most periodic functions can be represented to a reasonable approximation by the sum of the first few harmonics, so the power spectrum consists of peaks at just a few discrete frequencies.

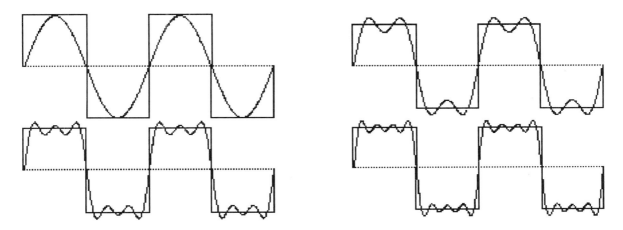

Fig. 11.58 Fourier series representation of a square wave

We saw that the spectrum for the square wave contained just a few significant peaks. We can understand this by examining the Fourier series for the square wave. (See Figure 11.58). It happens that the series representation includes only odd harmonics. In the figure, the upper left diagram shows the square wave approximated as a sine wave at the fundamental frequency. The upper right plot shows the fundamental plus a sine wave at three times the frequency. The bottom two plots add in sine waves at five times and seven times the fundamental frequency. You can see that the representation becomes increasingly accurate. Note that the amplitude of the harmonics decreases as the inverse of the frequency.

For our next example, we examine narrow-band noise, which is a random signal in which the signal oscillations occur in a relatively narrow frequency range. Such signals are often generated by physical systems that produce nearly periodic oscillations. An example is the flow of air over a normal cylinder. A train of vortices trail behind the cylinder as illustrated in Figure 11.59. The vortices shed at fairly regular but not precise intervals. A velocity sensor placed in the wake of the cylinder would record a signal that had most of its power in a narrow frequency band.

Our final example is broad-band noise (see Figure 11.60) which is usually produced by random processes. The particular signal in the figure was produced by a velocity sensor

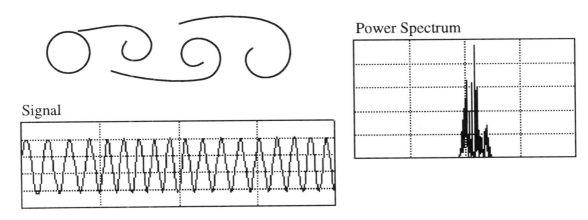

Fig. 11.59 Example of narrow band noise

152 Analysis of Sampled Data

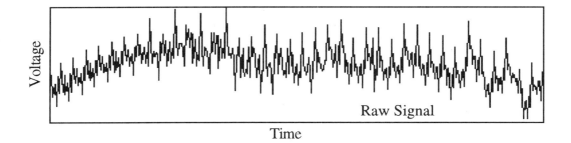

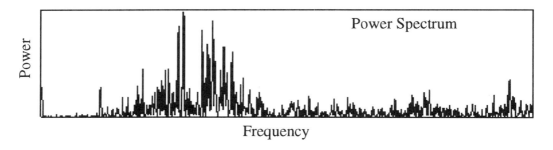

Fig. 11.60 Raw signal and power spectrum for broad-band noise

in a turbulent flow. The spectrum shows that the power in the signal is spread over a wide range of frequencies. It is not unusual, though, to still see a substantial peak in the spectrum as illustrated here.

Measurement of the Power Spectrum

The measurement procedures we will use here are all based on the fact that the power spectrum is the square of the Fourier transform of the original signal. Our normal procedure will be to sample the data at regular intervals, calculate the Fourier transform using a numerical procedure, and square the result to obtain the power spectrum. We will spend considerable time exploring problems that arise with this procedure. To understand these problems fully would require considerable mathematical analysis, which is not appropriate for this tutorial. The details are left to textbooks on the Fourier transform and digital signal processing.

We will first describe the Fourier transform and its numerical implementation. We will next discuss the properties of the power spectrum evaluated in this way. This will lead to an understanding of the data sampling criteria and of preprocessing of the data prior to transformation. We will end with the development of several VIs useful for spectral analysis of signals.

The Fourier transform is a transformation between the time domain and the frequency domain. That is, it transforms a function of time, x(t), into a function of frequency X(f). Formally, the Fourier transform is defined as:

$$X(f) = \int_{-\infty}^{\infty} x(t) e^{-i2\pi ft} dt.$$

where i is the square root of -1. The function X(f) is complex, which means that it has real and imaginary parts. We can understand this if we make use of the identity:

$$e^{-i\theta} = \cos\theta - i\sin\theta.$$

Plugging this into the definition of the Fourier transform we get:

$$X(f) = \int_{-\infty}^{\infty} [x(t)\cos 2\pi ft - ix(t)\sin 2\pi ft]dt.$$

The result can be split into real and imaginary parts:

$$X(f) = \phi(f) + i\psi(f)$$

$$\phi(f) = \int_{-\infty}^{\infty} x(t)\cos 2\pi ft\, dt; \qquad \psi(f) = \int_{-\infty}^{\infty} x(t)\sin 2\pi ft\, dt.$$

To help understand the equations above, we will next study the real part of the Fourier transform of a triangle wave. The original signal x(t) is a triangle wave at a frequency of 100 Hz. Now suppose we want to compute the Fourier transform of this signal for a frequency of 27 Hz, that is, we want ϕ(27 Hz). We form the product: x(t)cos2πft as shown in Figure 11.61, and integrate. We see that the product has both positive and negative values and the integral is small. Thus, the real part of the Fourier transform at this frequency is small.

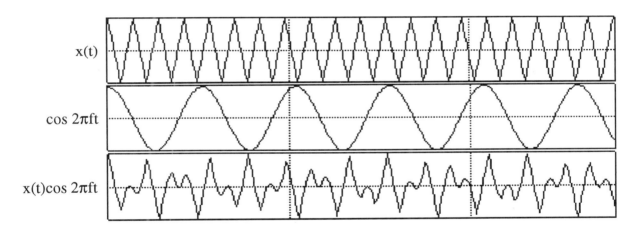

Fig. 11.61 Illustration of the Fourier transform calculation at f = 27 Hz for a 100 Hz triangle wave

Continuing the example, we now calculate the Fourier transform at the signal frequency, i.e., 100 Hz. (See Figure 11.62.) Now we see that when the triangle wave is positive the cosine is also positive. The product is always positive and the integral then has a large positive value. Figure 11.63 shows what happens when we compute the Fourier transform at a higher frequency. In this case, we look at the computation for a frequency of 160 Hz. You see that, once again, the product has positive and negative values, and the integral of the product is small. Finally, we look at what happens when we calculate the Fourier transform at a harmonic frequency. In Figure 11.64 we look at a frequency of 300 Hz, which is the strongest harmonic. There are both positive and negative values of the product, but positive values predominate, so the integral is positive. Summarizing, we see that $\phi(f)$ is small or zero at f = 27 and 160 Hz. It is large at 100 Hz and has an intermediate value at the harmonic frequency 300 Hz.

154 Analysis of Sampled Data

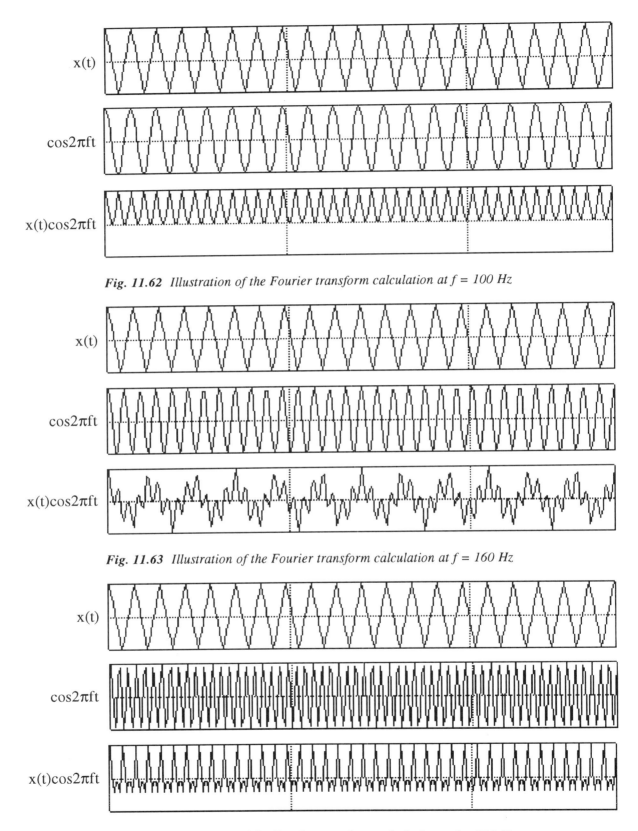

Fig. 11.62 Illustration of the Fourier transform calculation at $f = 100$ Hz

Fig. 11.63 Illustration of the Fourier transform calculation at $f = 160$ Hz

Fig. 11.64 Illustration of the Fourier transform calculation at $f = 300$ Hz

We see that the Fourier transform has useful properties; it tells us at which frequencies there is power in the signal. As we said earlier, the power spectrum is actually the square of the Fourier transform:

$$Gx(f) = \phi^2(f) + \psi^2(f).$$

There is a bit of a problem, though. The Fourier transform is finite only if:

$$\int_{-\infty}^{\infty} |x(t)dt| < \infty.$$

For stationary random data, this integral is always infinite. We get around this by defining a finite range Fourier transform as:

$$X(f,T) = \int_0^T x(t)e^{-i2\pi ft}dt.$$

This is more relevant to actual data, which we can sample only for a finite time. The restriction to a finite range does have an effect on the calculated power spectrum. We will ignore this for the time being, coming back to it after we discuss numerical calculation of the Fourier transform.

When we measure the power spectrum using a computer, we acquire a series of samples of the signal at regular time intervals. Assume that we have acquired a set of N samples with a sampling interval Δt between samples. The total sampling period is then $T = N\Delta t$. We approximate the Fourier transform integral as:

$$X(f,t) = \Delta t \sum_{n=0}^{N-1} x_n e^{-(i2\pi fn\Delta t)}.$$

where x_n is the ith sample and f is the frequency of interest. We normally calculate the transform at discrete frequency intervals that are multiples of the fundamental frequency, $f_o = 1/T$. The formula above becomes:

$$X(f_k,t) = \Delta t \sum_{n=0}^{N-1} x_n e^{-(i2\pi \frac{kn}{N})}.$$

The Discrete Fourier Transform is a good approximation of the true Fourier Transform as long as the sampling rate is fast enough to provide a good representation of the signal. Using discrete sampling does introduce a problem called aliasing, which we will discuss in the next section.

You can see that the calculation of the Discrete Fourier Transform could be very time consuming if the number of samples were large. The number of computational operations increases as the square of the number of samples, so the computation times quickly become prohibitive for long sample records. The Fast Fourier Transform (FFT) algorithm is a clever rearrangement of the operations that dramatically speeds up the calculation. The number of computational operations increases as N times the logarithm of N for the FFT, so it is considerably faster for long data records. Details of the algorithm are beyond the scope of this tutorial. The method is well covered in many texts on numerical analysis.

The Fast Fourier Transform is not an approximation of the Discrete Fourier Transform. It gives the identical results but is faster. The only effect of the FFT on the user is that the length of the sample record must be an integer power of 2. That is, you may use records of 2,048 or 4,096 samples but not 2,000 or 4,000. FFT algorithms are widely available and easily programmed. LabView has an FFT VI and also uses the FFT in its power spectrum VI.

Folding and Aliasing

Sampling at discrete time intervals restricts the maximum frequency that may be present in the signal; all frequencies must be less than half of the sampling frequency. Higher frequencies are not just ignored; they contaminate the measured spectrum at lower frequencies in a process called aliasing. Therefore, in measuring the spectrum of a signal, we must ensure that no frequencies greater than half of the sampling frequency are present.

To understand aliasing, we must first learn about folding. Folding means that the spectrum calculated using the discrete Fourier transform will appear to be folded around the Nyquist folding frequency, which is just equal to one half of the sampling frequency. Folding is illustrated in Figure 11.65, which shows the power spectrum for a signal sampled at 2,000 Hz. The Nyquist folding frequency is thus 1,000 Hz. The upper plot shows a broad-band spectrum calculated for frequencies ranging from zero up to the sampling frequency. We see that the spectrum appears to be folded about the Nyquist frequency. It is important to note that the original signal is band-limited; all of the power is at frequencies below 1,000 Hz. Therefore, the folding doesn't create any confusion in interpreting the spectrum. If we were to continue to calculate the Fourier transform at higher and higher frequencies using the same sampled data record, we would find the power spectrum repeating itself periodically as illustrated in the lower plot. No additional information is gained in performing this computation.

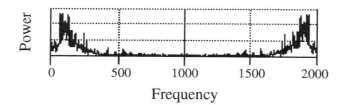

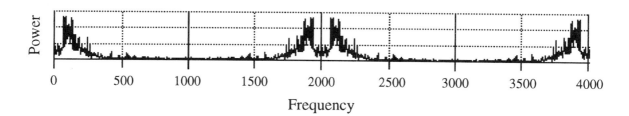

Fig. 11.65 Illustration of folding of a spectrum

We illustrate the cause of folding by an example. In Figure 11.66, the upper picture shows a 400-Hz sine wave sampled at 700 Hz. If we calculated the discrete Fourier transform for a frequency of 400 Hz, we would see a strong peak. The lower picture shows that the same set of samples could also represent a sine wave at 300 Hz. Therefore, calculation of the discrete Fourier transform at 300 Hz would also show a strong peak even though there is no 300 Hz component in the signal.

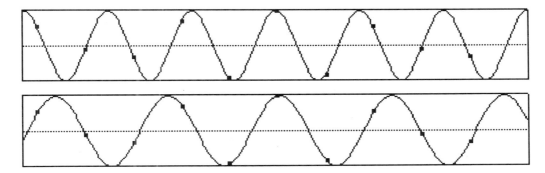

Fig. 11.66 Illustration of folding: The upper plot shows a 400 Hz sine wave sampled at 700 Hz. The lower plot shows that the same samples could be interpreted as representing a 300 Hz sine wave.

Aliasing occurs when the calculated spectrum is contaminated by folding. For example, imagine that the true spectrum of a signal is as sketched in the left half of Figure 11.67. The sampling is too slow, so there is significant power in the signal at frequencies above the Nyquist frequency. When the power spectrum is calculated, we find that the spectrum components above the Nyquist frequency are folded back into the calculated spectrum as illustrated on the right. The true spectrum cannot be extracted, because the sampling was too slow. A spectrum that turns up at high frequency is a characteristic warning sign of aliasing. You should learn to recognize it.

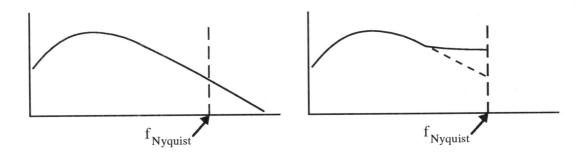

Fig. 11.67 Aliasing of a spectrum that was sampled too slowly

There are two approaches to avoiding aliasing. The first is to double the sampling rate and check for any change in the measured power spectrum. This is a time-consuming procedure that does not always work well. For example, assume that the true spectrum looks like the one sketched in Figure 11.68. We see that most of the power is in frequencies below 250 Hz. However, there is a small amount of power out to much higher frequencies. In many cases, our only interest would be in the region around the peak. We would like to resolve this part of the spectrum as accurately as possible. The best solution in this case is to low-pass filter the signal with cutoff frequency of around 400 Hz. This would eliminate the high-frequency portions of the signal, which may contaminate the spectrum via aliasing. You should use a low-pass filter whenever you are measuring the power spectrum to ensure that you do not have a problem with aliasing. An appropriate block diagram is shown in Figure 11.69.

158 Analysis of Sampled Data

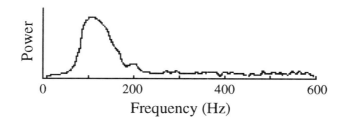

Fig. 11.68 Spectrum with a long tail at high frequency

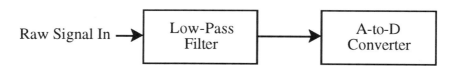

Fig. 11.69 Appropriate use of an anti-aliasing filter

We now illustrate aliasing with a couple of simple examples. In the first, the signal is a pure sine wave at 1000 Hz. The plot on the left of Figure 11.70 shows the measured power spectrum when the signal is sampled at 2,500 Hz. A peak is seen at the expected 1,000 Hz. The plot on the right shows the measured spectrum when the same signal is sampled at 1,400 Hz. The peak now appears at 400 Hz. The 1,000-Hz peak was reflected about the Nyquist folding frequency of 700 Hz. If we calculated the spectrum out to 1,400 Hz, we would see the peak at 1,000 Hz. However, we would be unable to tell which was the true frequency of the signal. Therefore, aliasing would render our measured spectrum useless.

The next example of aliasing is a little trickier. In this case, the signal is a square wave at 1,000 Hz. We expect a peak in the spectrum at 1,000 Hz with significant harmonics at 3,000 and 5,000 Hz. The plot in Figure 11.71 shows the measured spectrum when the 1,000 Hz square wave is sampled at 5,000 Hz. The spectrum is folded about the Nyquist frequency of 2,500 Hz. The fundamental peak shows up correctly at 1,000 Hz, but the 3,000-Hz harmonic now shows up at 2,000 Hz, and the 5,000 Hz harmonic shows up at 0 frequency.

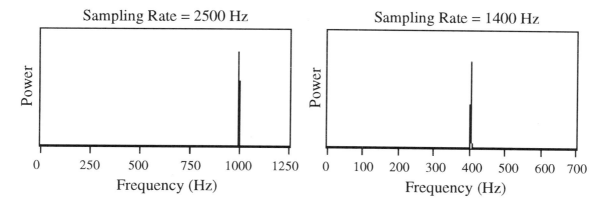

Fig. 11.70 Example of aliasing in measurement of sine wave frequency

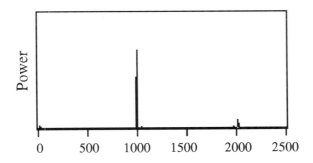

Fig. 11.71 Aliased spectrum showing displaced harmonics

Frequency Resolution and Leakage

Recall that the Fourier transform is an integral over all time. This integral cannot be calculated with a finite length record, so we defined a finite-time Fourier transform as the integral over a finite time interval 0->T. We now must examine what this does to our measured spectrum.

The finite-time Fourier transform is the equivalent of the full Fourier transform of the signal multiplied by a window function ß(t), as illustrated in Figure 11.72. The particular window function called a boxcar is zero everywhere except in the interval 0->T, where it is unity. We assume that the Fourier transform of x(t) would yield the true power spectrum. We analyze the effect of finite sample size by analyzing the Fourier transform of the product as illustrated on the figure.

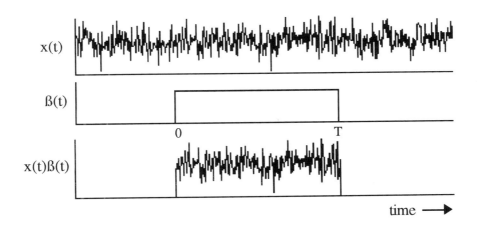

Fig. 11.72 The raw signal, x(t) multiplied by the boxcar function, ß(t)

The theory of Fourier transforms tells us that the multiplication of two functions in the time domain is equivalent to a convolution of the two functions in the frequency domain. That is, the Fourier transform of the product of two functions is equal to the convolution of the Fourier transforms of the two separate functions:

$$X[g(t)h(t)] = G(f) * H(f)$$

where H(f) and G(f) are the Fourier transforms of h(t) and g(t) and the asterisk means convolution. The convolution is defined as:

$$H(f) * G(f) = \int_{-\infty}^{\infty} H(u)G(f-u)du.$$

where u is a dummy variable of integration. This is a pretty confusing expression. It is most easily understood by examining the convolution at a single frequency f_o:

$$H * G \big|_{f_0} = \int_{-\infty}^{\infty} H(u)G(f_0-u)du.$$

Let's look at an example to make the idea of a convolution a little clearer. We will talk about the convolution of two functions H(f) and G(f), shown at the top of Figure 11.73. Plotting G(fo-u) just shifts the function G(f) so that it is centered around the point fo. To form the convolution at the frequency f_0, we multiply the function H by the shifted function G and integrate over all u. To compute the full convolution, we have to repeat this process for all different possible values of f_0. We see that if G is a smooth, band-limited function as in our example, the convolution acts to smear out the bumps in the spectrum H(f).

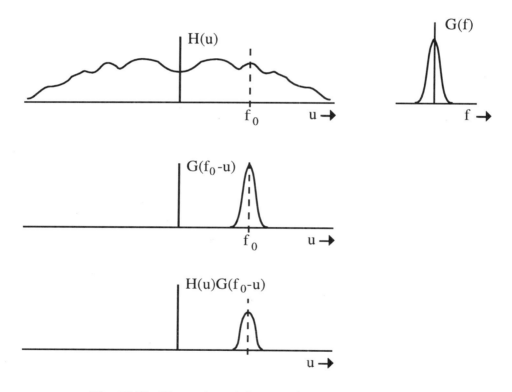

Fig. 11.73 Illustration of the convolution of two functions

The function G(f) in the previous example is fairly typical of the window functions that we will eventually use. The function is smooth and non-zero in only a relatively narrow frequency band. We can think of such a function as a filter applied to the true spectrum. That is, the convolution will act to smooth out any bumps in the true spectrum. Consider the case in which the original signal h(t) is just a sine wave so that H(f) is just a spike. Assume that the function G(f) is a smooth bump, as shown in Figure 11.74. The convolution will have non-zero values only in the range fo-Δf < f < fo+Δf, with a peak value at f = fo. In fact, the convolution will smear out the original spike to look just like the

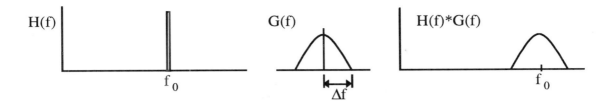

Fig. 11.74 Convolution of a pure tone spectrum with a smooth narrow-band filter function

function G(f) centered on fo. It is important to remember that we never actually perform the convolution calculation. We get this automatically when we Fourier transform a signal that has been multiplied by a window function.

It is simple to work out the effective filter function that is applied when we use a boxcar window function, ß(t). Assume that the total sampling period is T. Then we represent the boxcar function as:

$$\text{ß}(t) = \begin{cases} 1 & \text{if } -1/2 T < t < 1/2 T \\ 0 & \text{otherwise} \end{cases}$$

We calculate the Fourier transform as:

$$B(f) = \int_{-\infty}^{\infty} \text{ß}(t) e^{-i2\pi f t} dt = \int_{-.5T}^{.5T} e^{-i2\pi f t} dt = \frac{1}{\pi f} \sin \pi f t.$$

The function B(f) is called the sinc function and is sketched in Figure 11.75.

The sinc function does not look like a very nice filter function to apply to our spectrum. The main peak is reasonably narrow. In fact, we will see that the central peak is narrowest for a boxcar window function. However, the oscillating tails, called sidelobes, are a big problem. They will act to smear the spectrum out considerably.

If we ignore the sidelobes, we can estimate the bandwidth of the effective filter as the width of the central peak, 2/T. This gives us a rough estimate of the measurement resolution possible. For example, if we acquired a set of 2,048 samples at 10,000 samples/second, our sampling period T would be 0.2048 seconds and the frequency resolution would be 2/0.2048sec = 9.77 Hz. The spectrum is calculated at frequency increments of 1/T= 4.88 Hz. If two peaks in the true spectrum were separated by 2 Hz, we could not resolve them. They would appear as a single broad peak in the measured spectrum.

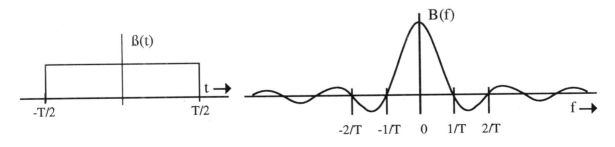

Fig. 11.75 The boxcar and its Fourier transform the sinc function

162 Analysis of Sampled Data

A filter function that does not decay monotonically is disturbing. A smooth filter seems more appropriate. Also, the sidelobes further reduce our resolution so that it is actually considerably worse than 2/T. This effect is called leakage and is illustrated in our next example. The three spectra shown in Figure 11.76 were all measured from the same signal, a pure 100-Hz sine wave. The sampling frequency was 300 Hz for each case. A set of 128 samples was used for the left plot, giving a sampling period of 0.43 seconds and a resolution of 4.7 Hz. A set of 256 samples was used for the middle plot, so the resolution was 2.3 Hz. Finally, 1,024 samples were used for the right plot, giving a sample period of 3.4 sec and a frequency resolution of 0.59 Hz. We see that the estimate of 2/T as the frequency resolution is only approximate. The width of the measured spectral peaks is actually somewhat greater than that in these cases because of the contribution of the sidelobes.

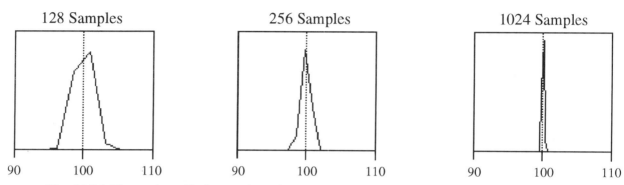

Fig. 11.76 Illustration of leakage, which decreases with increasing sample record length

Alternative Window Functions

Generally, we prefer an effective filter function whose amplitude decreases monotonically from the peak value. We can change the effective filter function by selecting a different window function whose Fourier transform is more well behaved. A commonly used window function is the triangle shaped Parzen window sketched in Figure 11.77. Its Fourier transform shows a main peak that is wider than the sinc function, but the sidelobes are considerably weaker. Therefore, leakage is reduced, and it is much easier to estimate the frequency resolution using a Parzen window than the boxcar window.

Another useful window function is the Hanning window represented by the equation:

ß(t) = 1 - cos(2πt/T).

The Fourier transform of the Hanning window has a wider central peak than either the boxcar or Parzen windows but the sidelobes are very weak making this an excellent effective filter to apply to a spectrum. (See Figure 11.78.) The plot shows that the

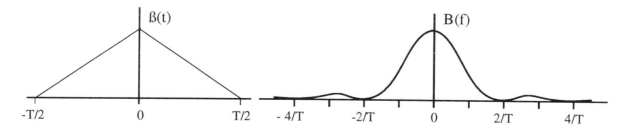

Fig. 11.77 The Parzen window (left) and its Fourier transform (right)

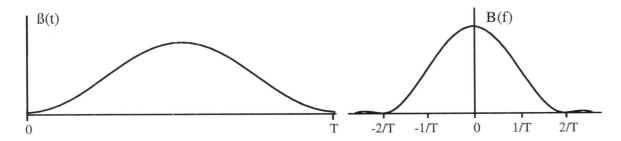

Fig. 11.78 The Hanning window (left) and its Fourier transform (right)

frequency resolution of the measured spectrum is about three to four times 1/T when the Hanning window is used. This resolution estimate should be used to set up the sampling parameters. The desired frequency resolution should be determined first, then the sampling rate and the number of samples selected to give an appropriate sampling period, T.

What exactly does it mean to apply a window function that differs from the standard boxcar function? Each data sample must be multiplied by a weight function which depends on position in the data record. For example, to implement the Parzen window, each sample is multiplied by the weight:

$$w_i = 1 - \left| \frac{i - \frac{1}{2}(N-1)}{\frac{1}{2}(N+1)} \right|$$

where w_i is the weight and N is the total number of samples. The Hanning window is:

$$w_i = \frac{1}{2}\left[1 - \cos\left(\frac{2\pi i}{N-1}\right) \right].$$

These windows can be applied automatically as LabView functions. Several other smoothing windows are also available in LabView.

Statistical Convergence

Just as in the measurement of the mean and the standard deviation, we must be concerned about the statistical variations in our estimates of the power spectrum. We start with an example to illustrate this. Consider a case in which we are measuring the spectrum illustrated in Figure 11.79. We see that there is a peak in the spectrum around 50 Hz and significant power up to a frequency of 500 Hz. A sampling frequency of 1 kHz is selected to avoid aliasing, and 2,048 samples are used to get a frequency resolution around 1 to 2 Hz. The problem is that there would not be a very good statistical sample of the lower frequency portions of the signal. The signal is dominated by fluctuations around 50 Hz, and only about 100 cycles of this fluctuation would occur in the 2.048 second sampling period. The measured spectrum would appear very rough in the low-frequency range, and if the measurement were repeated, the spectrum would not look the same. It would be impossible to discern the true shape and height of the spectral peak.

The answer to the problem of inadequate statistical convergence is to increase the length of the sample record. However, there are right and wrong ways to do this. The wrong way is to increase the number of points used in your spectrum measurement. The computation time increases rapidly, and you can run into memory limitations on your computer. The number of points used should be determined only by the desired frequency resolution.

164 Analysis of Sampled Data

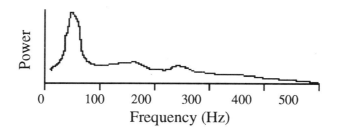

Fig. 11.79 Spectrum plot illustrating problems of statistical convergence

The appropriate way to reduce statistical variations is to repeat the measurements and average the resulting spectra. The simplest way to do this is to acquire multiple sample records, calculate the power spectrum from each, and then average the results. This way is the most efficient in terms of total time required if the time required is dominated by the time it takes to perform the calculations. This would be the case for relatively high frequency signals where the sample record is acquired very rapidly. However, if the total time is dominated by the time it takes to acquire the sample records, there is a more efficient technique which is discussed below.

The alternative method for calculating averaged spectra is to acquire a very long sample record, then partition it into overlapping segments, as indicated in Figure 11.80. Returning to our previous example, we might acquire a sample record of 8 X 2,048 = 16,384 samples. This would then be divided into 16 overlapping segments, each containing 2,048 samples. The first segment would include samples 0 to 2,047 and the second segment would contain samples 1,024 to 3,071. Once again, the power spectrum would be calculated for each segment and the results averaged. This technique appears to use the same data twice. However, the statistical uncertainty will be considerably less than if the record were simply divided into nonoverlapping segments. This technique is preferred when the actual data acquisition is time-consuming.

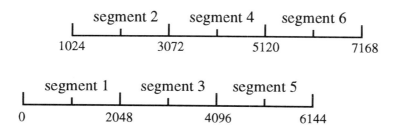

Fig. 11.80 Segmenting of the data record to get better statistical convergence from a fixed length record

Measurement of the Power Spectrum using LabView

We now develop a set of VIs for measuring the power spectrum of a voltage signal that we periodically sample using the A-to-D converter. You could develop similar VIs for data sampled through other channels, such as the parallel digital input port.

The basic steps in measuring the spectrum are i) acquire the array of data samples, ii) apply an appropriate window function, iii) calculate the Fourier transform using an FFT routine, and iv) square the Fourier transform to obtain the power spectrum. LabView has several built-in functions that make this easy. Several different window functions are available. We use the Hanning window. When the data array is connected to the Hanning icon, the array is automatically multiplied by the window function. LabView also has a Power Spectrum VI that performs the FFT of the data array and squares the results to get the power spectrum. The Power Spectrum VI calculates the spectrum at frequencies up to the sampling frequency. The second half of this range is redundant; the results are just a reflection of the first half. You will see that we eliminate the second half of the results array in the LabTutor Spectrum VIs.

Three VIs are shown below. The first, called Spectrum, acquires the data through the ADC, subtracts the mean value from each voltage sample, applies a Hanning window, calculates the power spectrum, and normalizes the spectrum by the signal variance. This VI is convenient to use for quick measurement of a spectrum. It can be used as a subVI in more complicated instruments you may develop. The next VI, called SpecPlot, is the same as Spectrum except that the data are displayed as a plot.

The final VI, called SpecAve, is the one you should use for careful measurements of a spectrum. This VI is used to repeatedly measure the power spectrum of a signal and average the results. This will smooth out the statistical variations that occur when you calculate the spectrum using a single sample record. The user should increase the number of cycles used until the power spectrum appears smooth.

Figure 11.81 shows the front panel for Spectrum. It contains the usual sampling controls, including Channel, Number of Scans, Scan Rate, and Input Limits. The Iteration and Clear Acquisition controls should be set to the values shown here when the VI is not used in a loop. When the VI is used in a loop, wire the Iteration input to the Loop Iteration terminal, i. The Clear Acquisition input must have a True value on the last iteration of the loop. The number of samples must be a power of two. The raw voltage data are available in Output Array and the power spectrum in another array.

The diagram is shown in Figure 11.82. The first step is acquisition of the data. The AI Waveform Scan VI is used to acquire the data through the A-to-D converter. The output of AI Waveform Scan is a two-dimensional array, even though you acquire samples from only one channel. The Index Array function is used to select the column that contains the data record, and the data are displayed on the front panel. The mean and the standard deviation of the data samples are calculated, and the mean is subtracted from each data sample before sending the data array to frame 1. The square of the standard deviation is calculated also, since it will be used to normalize the spectrum results. In the second frame

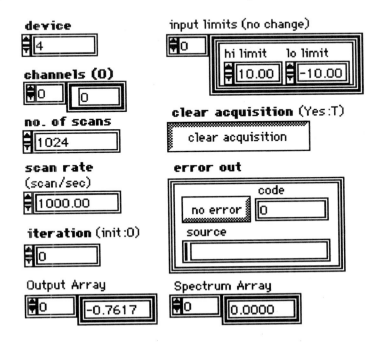

Fig. 11.81 Front panel of the Spectrum VI

166 Analysis of Sampled Data

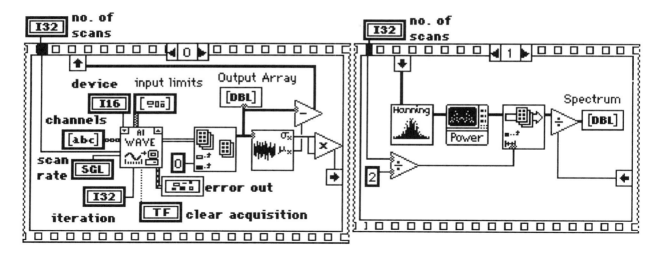

Fig. 11.82 Diagram for the Spectrum VI

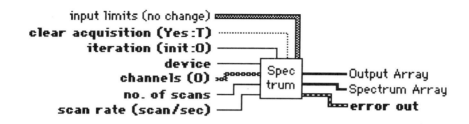

Fig. 11.83 Icon for the Spectrum VI

frame, the Hanning window is applied to the array of data. The power spectrum is calculated using LabView's Power Spectrum VI. This VI produces values up to the sampling frequency. Only the first half of these are useful, so the second half of the array is truncated. The results are divided by the variance and displayed on the front panel. We will use the Spectrum VI as a subVI in the more complex VIs to follow, so its icon and connections are shown in Figure 11.83.

The front panel for SpecPlot is the same as Spectrum, with the addition of the plot (see Figure 11.84). The number of samples must be a power of two. The power spectrum is displayed as a plot, with the axes automatically scaled for the data sampling parameters. You may want to change the frequency axis to a logarithmic scale. This is easily done with the LabView front panel controls.

The heart of this VI is the Spectrum VI we just developed as seen in Figure 11.85. The output array from Spectrum is sent directly to the front panel plot. The Fourier transform is calculated on intervals of $\Delta f = 1/T$, where T is the total sample period. The total sample period is just equal to the Number of Scans divided by the Scan Rate. Therefore, the frequency interval for the plot is the Scan Rate divided by the Number of Scans. This frequency increment is supplied to the graph. You can adjust the scale on the front panel to make it look nice.

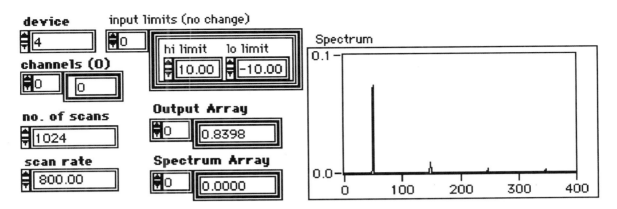

Fig. 11.84 Front panel for the SpecPlot VI.

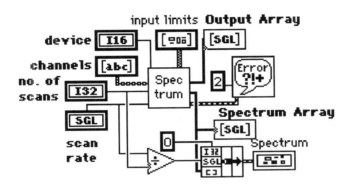

Fig. 11.85 Diagram for the SpecPlot VI

The SpecAve VI runs the Spectrum VI a specified number of cycles and averages the power spectra. The controls are the same as those for SpecPlot, except for the addition of the control for the number of cycles. (See Figure 11.86.) The number of samples should be chosen to give the frequency resolution desired. The number of cycles should be large enough to give a smooth spectrum.

Frame 0 of the diagram (Figure 11.87) shows the Spectrum VI inside a For Loop. Each time the loop executes, a new spectrum is calculated. Thus, the result of the loop is a two-dimensional array containing N power spectra, where N is the number of cycles. Note how we have used the Iteration and the Clear Acquisition terminals of the Spectrum VI. The Iteration terminal is wired to the Loop Iteration terminal, i. A True value must be sent to the Clear Acquisition terminal of the Spectrum VI on the last iteration of the For Loop. This is obtained by comparing the desired number of loop cycles to the value in the Loop Iteration terminal.

The power spectrum array from frame 0 is transposed using the Transpose 2D Array function. This is done so that the power spectrum for each cycle is in a row and the mean can be calculated for each frequency in the For Loop. The result is plotted on the front panel, with the frequency axis determined in the same way as in SpecPlot.

168 Analysis of Sampled Data

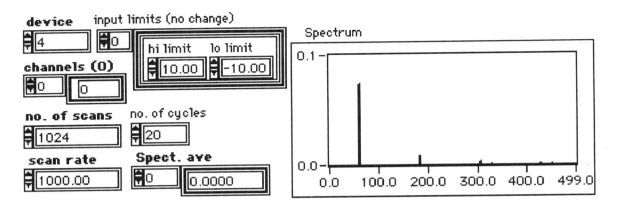

Fig. 11.86 Front panel for the SpecAve VI

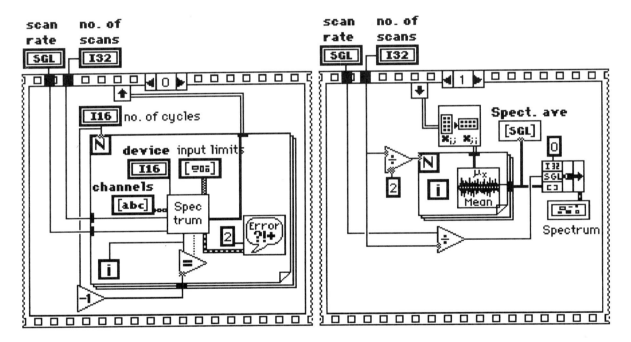

Fig. 11.87 Diagram for the SpecAve VI

Index to the LabTutor VIs

ADCExer3	59
ADDB	70
ADDBTrig	73
AO Write One Update Control Loop2	78
AutoCorr	144
Control Loop	76
Control Loop2	78
Corr	131
Corr2	132
CorrFun	136
DACExer2	43
DACExer3	45
DACExer3a	45
DBGraph	71
DigExer3	90
DigExer4	90
DM5120	101
Fluke	99
MeanComp	121
MeanComp1	122
MeanComp2	123
PDF	110
PlotFile	69
RunAve	116
SpecAve	165
SpecPlot	164
Spectrum	163
TestDatPlot	112
Uncer	125
Vave	64
Vsave	67
VseqP	61
VseqP2	65